A Biblical Case
for an
Old Earth

A Biblical Case
for an
Old Earth

DAVID SNOKE

BakerBooks
Grand Rapids, Michigan

© 2006 by David Snoke

Published by Baker Books
a division of Baker Publishing Group
P.O. Box 6287, Grand Rapids, MI 49516-6287
www.bakerbooks.com

Printed in the United States of America

All rights reserved. No part of this publication may be reproduced, stored in a retrieval system, or transmitted in any form or by any means—for example, electronic, photocopy, recording—without the prior written permission of the publisher. The only exception is brief quotations in printed reviews.

Library of Congress Cataloging-in-Publication Data
Snoke, David, 1961–
   A biblical case for an old earth / David Snoke.
      p.   cm.
   Includes bibliographical references.
   ISBN 10: 0-8010-6619-0 (pbk.)
   ISBN 978-0-8010-6619-1 (pbk.)
   1. Bible and geology. 2. Earth—Age. I. Title.
BS657.S66 2006
231.7′652—dc22   2006009543

Unless otherwise indicated, Scripture is taken from The Holy Bible, English Standard Version, copyright © 2001 by Crossway Bibles, a division of Good News Publishers. Used by permission. All rights reserved.

Scripture marked NIV is taken from the HOLY BIBLE, NEW INTERNATIONAL VERSION®. NIV®. Copyright © 1973, 1978, 1984 by International Bible Society. Used by permission of Zondervan. All rights reserved.

Scripture marked NASB is taken from the New American Standard Bible®, Copyright © 1960, 1962, 1963, 1968, 1971, 1972, 1973, 1975, 1977, 1995 by The Lockman Foundation. Used by permission.

The quote on p. 136 is from *The Magician's Nephew* by C. S. Lewis copyright © C. S. Lewis Pte. Ltd. 1955. Extract reprinted by permission.

Contents

Preface   7

1. Starting Assumptions   11
2. The Scientific Case   24
3. The Biblical Case I: Animal Death   47
4. The Biblical Case II: The Balance Theme in Scripture   76
5. The Biblical Case III: The Sabbath   99
6. Concordantist Science   114
7. Interpreting Genesis 1 and 2   132
8. The Flood of Noah   158
9. Implications for Theology   176

Appendix: A Literal Translation of Genesis   197
Notes   213
Subject Index   218
Scripture Index   221

Preface

This book was instigated by a debate within the Presbyterian Church of America (PCA), my own denomination, in regard to the orthodoxy of the old-earth position. "Young earthers" say that the earth and all of creation is at most ten or twenty thousand years old, essentially the same age as the history of modern humans. "Old earthers" say that the earth is billions of years old, in agreement with the assumptions of modern geology. Some Christians insist that the old-earth position is theologically heretical, or at least heterodox, and some in my denomination want to deny pastors the right to preach if they do not hold to a young-earth view. The debate is not restricted to my denomination, however. Unfortunately, this issue threatens to divide Christians—many well-known seminary professors and teachers such as Meredith Kline and Michael Horton adhere to an old-earth view, while notable figures such as John MacArthur, for whom I have great respect, have publicly called the old-earth position theologically "liberal," or heterodox.

Theological liberalism does exist. We have seen a century of slide on almost every church doctrine. Because many Christians react so strongly against liberalism, however, sometimes leveling the charge of "liberalism" is an easy way to dismiss an opposing argument. Some Baptists dismiss those who baptize infants as liberal; Catholics who believe in the Latin rite dismiss those who believe in using the vernacular as liberal. This happens with more esoteric issues, as well: some people who believe in the eschatological doctrine of a pre-tribulational Rapture view all who disagree as liberal; within my own denomination, some have dismissed Presbyterian authors like R. C. Sproul and Francis Schaeffer as liberal because their writings do not conform to the doctrines of presuppositional apologetics expounded by Cornelius van Til. On many of these issues, however, Christians have learned that we can fellowship with people with whom we disagree on broad issues of interpretation of Scripture because we know that at least they share with us a strong view of the inerrancy and primacy of Scripture. Each of us must be convinced from Scripture, and often this means we must adopt a minority view for the sake of conscience, even if most conservatives believe otherwise.

In this book I argue that the old-earth position is a valid, conservative, and orthodox interpretation of the Bible. This may shock some people—the young-earth position is so equated with orthodoxy that when I say that I believe in an old earth, people have sincerely asked me if I also deny the virgin birth, the bodily resurrection of Jesus, etc. This is partly because theological liberals assume that the earth is old without even a debate and mock the young-earth position, so that people associate the old-earth view with theological liberalism. Perhaps, however, people also make this association because some who adhere to the young-earth view encourage it, thereby preventing their opponents from getting a fair hearing among conservatives.

In many people's eyes, I have probably lost before I begin, because no matter what I argue from the Bible, they will say, "But you have come up with this just because you want the Bible to agree with science." I freely confess to this charge—I would not have studied this issue with as great interest had I not wanted to see whether a young-earth view was strictly necessary. I discuss the validity of such an approach in chapter 1. I hope by the end, however, that if readers have not been persuaded to agree with my views, they will at least agree that my arguments are *biblical*, a viable position in a debate among Christians (similar to that over infant baptism) and not wild-eyed mangling of the Scriptures.

Many of the proponents of the old-earth view adhere to a "framework" model of interpretation of Genesis 1; the framework model takes this chapter as essentially poetic, not giving any chronological information. This book presents the case for a "day-age" view that takes Genesis 1 as giving a real chronological sequence, but not necessarily of twenty-four-hour days. This position is too quickly dismissed by both sides, although many Christians who are trained scientists, such as Hugh Ross and Robert Newman, find this view very appealing.

I thank Bruce Rathbun, Michael Schuelke, and George Hunter for critical reading of this manuscript in its early stages, and for many profitable discussions, despite our disagreements.

Starting Assumptions 1

At the very outset, let me say that my experience in science has affected my interpretation of the Bible. For some people, this is a cardinal sin. This is one of the most important issues before us. Is it ever permissible to allow our experience to affect our interpretation of the Bible? Or should I strive to study the Bible in an interpretive "vacuum," with no reference to any of my life experience? Is that possible?

To put it another way, it is very improbable that I ever would have come up with the view that the earth is millions of years old if I had never studied science. If I had never studied science, I also probably would not have come up with the idea that everything is composed of electrons, protons, and neutrons, or that life is based on DNA, or many other things that I believe. At first blush, one might say, "So what?" There are many things about the natural order that God has seen fit to let us discover by experience, which he does not discuss in the Bible. Most people have no trouble affirming that the theories of electrons, protons, and DNA, while not in the Bible, are compatible with the Bible.

The difference between an old-earth view and the theory of electrons is that the Bible talks directly about the origin of the world in several places, while it does not talk much about the composition of things. In saying that the earth is millions of years old, and at the same time saying that I affirm the Bible is true and has no errors, I must argue that those places where the Bible speaks on origins are compatible with belief in an old earth. With electrons and protons, I do not have to compare my theories to any particular Bible passages.

I believe that an old-earth view is compatible with the Bible. Nevertheless, I admit that my interpretation is a "possible" one, not an "obvious" one. The question that lies before us is therefore, "Is it ever legitimate to prefer a 'possible' interpretation over a simpler, 'obvious' interpretation, based on our experience?" I will argue that it is often legitimate.

Before making this argument, I want to make clear that there are also *illegitimate* ways of letting experience affect our interpretation of the Bible. First, some people might argue that when I say that "science" has affected my interpretation, what I really mean is that "peer pressure" has affected my interpretation. In other words, they might say that I have changed my interpretation of the Bible because it is unpopular among my colleagues. If they were right, my approach would be illegitimate, but I hope that they are wrong, and that I have not caved in to peer pressure. I will not argue that there have been no Christians who have capitulated to prevailing views for social reasons, nor will I argue that there is no anti-Christian social pressure among scientists. Many Christians in the sciences have capitulated, in my opinion, and many of us are familiar with the intolerant spirit of political correctness on our campuses, as expressed in such things as speech codes.

Some of us are familiar with pressure from the right, as well—from "fundamentalists" who use peer pressure to insist that anyone who rejects their view is "liberal." All of us, including scientists, need to carefully distinguish between what are really facts, and what is merely the opinion of a lot of people. It is illegitimate to change our view of the Bible because we want a more popular interpretation.

At the same time, we should recognize that no view that says that the Bible is true will be popular in the modern scientific world. In recent years, many scientists have joined together to form the "intelligent design" movement, which questions the mechanisms of Darwinian evolution while remaining open about the age of the earth. This movement has met with scorn in the secular academic world, as "creationism in a cheap tuxedo."[1] I hope that people will recognize the courageous stand that these scientists are taking.

Second, I want to make clear that science is not some kind of higher authority that trumps the Bible. Because something has been observed through a microscope or a telescope does not make it more important than Scripture. Science is just a way of expanding and organizing our experience; therefore, science has the same authority as any human experience. It is illegitimate to place anything generated by human beings in a position of unquestioned authority over the Bible.

The issue before us, then, is whether experience, including the expanded experience of science, may ever legitimately affect our interpretation of the Bible. I argue that honesty demands it in many cases.

The Moving Earth

For an example, let's first consider the old debate that involved Galileo, of whether the earth moves. Many young-earth advocates resent being compared to Ptolemaics, because these days no one seriously argues that the earth does not move. In the sixteenth century, however, many intelligent people did argue that the earth does not move. Luther, Calvin, and Melanchthon all rejected the idea of a moving earth—Luther is quoted as saying of Copernicus, "This fool wishes to reverse the entire science of astronomy, but sacred Scripture tells us that Joshua commanded the sun to stand still, and not the earth"; even John Wesley many years later said that the Copernican system "tends to infidelity."[2]

These people were not stupid. On one hand, there was little scientific evidence for a moving, round earth, and that scientific evidence was available only to a few experts; on the other hand, there were apparently sound biblical arguments that the earth does not move. Psalms 93:1, 96:10, and 104:5 say in very definite terms "The earth is firmly established; it can not be moved." What has changed since then—why doesn't anyone debate whether the earth moves any more? Primarily, our experience has changed. Many of us have the experience of changing time zones and seeing the sun rise at an earlier time, perhaps even flying around the world, looking at planets through a telescope, seeing pictures of the earth from outer space, etc. Perhaps the most famous historical demonstration of the earth's motion is Foucault's pendulum, which showed that an object in linear motion, with no sideways forces, rotates relative to us as the earth moves.

It seems obvious to us now that passages like Psalm 93:1 are poetic, referring to God's protection and maintenance of the earth, and not meant to imply that the earth does not rotate. Even the passage quoted by Luther, Joshua 10:12–13, in which the sun stands still, does not change the opinion of most Christians that the normal behavior of the earth is to move. Various explanations of this miracle have been proposed, such as a complete, miraculous suspension of the laws of momentum to stop the earth in its orbit, or more mundanely, a miraculous optical effect that kept the image of the sun in the same position in the sky even as the earth continued to rotate to accomplish the purpose stated in the text, namely to give Joshua more light. But in the sixteenth century these would not have been obvious interpretations at all! A critic might have said, "If you allow Psalm 93:1 to be taken as 'poetic,' why not other passages in the Psalms, for example Psalm 2:7–8, which teaches that God has a Son, or Psalm 32:1, which teaches that we can have our sins forgiven? Aren't we on a slippery slope to allowing us to reinterpret anything in the Psalms we don't like?"

The answer to this criticism is that many scholars have defined clear guidelines for interpretation of the Psalms to help us discern the meaning of the symbolism. Perhaps sometimes we err and "explain away" a passage in the Psalms that we should take more literally, but this does not mean that reading some things in the Psalms as symbolic or poetic makes interpretation of the Psalms entirely subjective. A sixteenth-century critic might answer, however, that our interpretation of the Psalms is suspect because no one ever would have deduced a moving, round earth from reading the Psalms. We have allowed our experience (in the sixteenth century, this would have been the experience of just a few elite scientists with telescopes) to affect our interpretation of the Bible.

Note that the Roman Catholic Church at the time of Galileo was perfectly willing to admit that the earth "appeared" to move.[3] It simply insisted, on the basis of the most natural interpretation of Psalm 93:1, etc., that the earth does not, in fact, move. Galileo could have published anything that said that the earth "appeared" to move, or that the simplest mathematical theory invoked the "useful fiction" of a moving earth. It was when Galileo moved from scientist to philosopher that he got into trouble, by insisting that the earth "really" moves.

Even today, one could change one's physics to put the earth in a fixed, non-moving reference frame. In that case, however, one would need to invoke numerous new forces to explain the Coriolis and centrifugal forces that occur in such a system and which give rise to things such as hurricanes. The observation that planets seem to move backward in the sky at times, known as retrograde, would also need an explanation. At some point, the sheer complexity of such a system leads one to say that if the earth does not "really" move, then God is a great deceiver to have made an entire universe that is perfectly harmonized to make the earth look like it moves, when it does not. One is faced with either utter complexity in analyzing experience, or a relatively simple change to a "possible" interpretation of the Bible.

Modern Apostles

Another area where our experience enters into Bible interpretation is in evaluating the claims of modern persons to be apostles like Peter and Paul. Many passages of the Bible can be used to make a very convincing case that the role of apostle should be a normal office of the Christian church, for example, Ephesians 4:11, which says, "It was he who gave some to be apostles, some to be prophets, some to be evangelists, and some to be pastors and teachers" (see also 1 Cor. 12:28). There is no explicit teaching in the New Testament that the role of apostle will cease.

Why then do most Christians believe this role has ceased to exist? Second Corinthians 12:12 says that the marks of an apostle are "signs, wonders and miracles."

Many Christians, however, take the view that signs and wonders have passed away; they take this from passages such as 1 Corinthians 13:8, "But where there are prophecies, they will cease." Such views are far from obvious or natural to many readers of the Scriptures—one must search high and low for oblique, vague references to this "passing away" of apostles and their signs and wonders. Much of the reason is that, in our experience, there is little good evidence for powerful signs and wonders of the type recorded in the Bible as accompanying the apostles (these are not the same as "answers to prayer," which many Christians experience in the normal course of life).

Biblical evidence for a passing away of signs and wonders is not completely lacking. Hebrews 1:1–2 says, "In the past God spoke to our forefathers through the prophets at many times and in various ways, but in these last days he has spoken to us by his Son, whom he appointed heir of all things, and through whom he made the universe." There is a sense of finality in this passage, that Jesus is the final word to humankind. Matthew 12:39 says, "A wicked and adulterous generation asks for a miraculous sign! But none will be given it except the sign of the prophet Jonah." Taking the word "generation" to refer to the race of the Jews, as it appears to mean in other contexts (e.g., Matt. 24:34–35), this would mean that the Jews were not to expect a continuing series of miracle workers like the Old Testament prophets. In Matthew 21:33–39, Jesus tells the parable of the tenants in the vineyard, in which he says that the "last" prophet sent to Israel will be the Son. Daniel 9:24 says that the coming of the Messiah will "seal up vision and prophecy." Taking these together, one can argue that Jesus is God's final message, and the apostles were accredited to act as his agents for one generation (see, for example, Matt. 10:1 and John 16:12–13).

Many Christians date the passing away of the apostolic age, along with its signs and wonders, to the destruction of the Temple in 70 AD. Those who believe in modern apostles, however, accuse those who don't believe in modern apostles of letting their experience influence their interpretation of the Bible. Clearly, if there were many credible apostles performing believable signs and wonders, the view that these offices still exist would have greater credibility. Many charismatics point to various miracle workers such as Oral Roberts as new apostles and prophets. Other Christians, including myself, just are not convinced by the evidence for real miracles done by these people. God still answers prayer, but the evidence that there are special people who can perform miracles on demand like the apostles' "Get up and walk!" is mostly lacking. Those who believe in modern apostles would say that we have started down the "slippery slope," however; if we will allow that the office of apostle has ceased, what about other roles in the New Testament, such as evangelist—has this too "faded away"? Those of us on the other side feel it is dangerous to accept all manner of claims without testing them against our experience.

Coming with Power

Another example of the tension between our experience and some biblical interpretations is the passage in Matthew 16:24-28 (parallel with Mark 8:34-9:1 and Luke 9:23-27) in which Jesus says (putting the three passages together as a composite), "there are some standing here who will not taste death until they see the Son of Man coming in the kingdom of God with power." The most "obvious" interpretation of this passage is that Jesus would return to judge the world within about forty years, the biblical length of time for a single generation to die off (Num. 14:32–33).

Why don't we believe this is what Jesus meant, then? Our experience tells us otherwise. We are still here, and Jesus has not judged the earth yet. We therefore take an alternate, "possible" interpretation of this passage. I believe that Jesus was referring to the destruction of Jerusalem in 70 AD when he referred to his "coming with power"; the judgment ready to fall on Jerusalem is a constant theme in the Gospels (Luke 21:20–24; Matt. 21:33–43; 23:37–39; cf. Mal. 4:5–6). Others who also believe in the inerrancy of Scripture have argued that Jesus referred to his transfiguration, or to his ascension, although in these cases it would seem obvious that not all standing there would die before then. In any case, none of these interpretations is obvious, though all are "possible." Theological liberals who reject the inerrancy of Scripture accuse us of "unnatural" interpretation—they wonder why we don't take the "plain meaning" and admit that these words of Jesus are false.

The King Who Was Not a King

A more obscure example is found in Daniel 5:1, which refers to "King Belshazzar." In the nineteenth century, Middle Eastern scholars scoffed at this reference as evidence of errors in the Bible, because other documents clearly showed that Belshazzar was not the ruler of the Babylonian empire at the time. More recent scholarship has shown that Belshazzar was a subordinate ruler, lord of the city of Babylon, not the ruler of the whole empire. A possible, reasonable interpretation of Daniel 5:1 is that the term "king" refers to this type of ruler, a viceroy, so to speak.[4] Should we reject this interpretation, which we clearly would never have arrived at without extra-biblical documents from the science of archaeology, and insist that Belshazzar had to be ruler of the empire?

In general, Christians have found great support for the Bible from archaeology and other scholarly study of the ancient Middle East. Sometimes those studies lead Christian scholars to revise our interpretation of the Bible, however. Interpretations that we might feel are obvious at first glance turn out to be wrong when we study the culture and the history more carefully. Extrabiblical experience and scholarship have brought about a change in our Bible interpretation.

Must We Interpret the Bible in a Vacuum?

All of these examples point to the legitimacy of allowing experience to affect our interpretation of the Bible. Why do we react against it, then? I have already mentioned two reasons. First, we rightly want to avoid any hint of concession to worldly views due to societal pressure. Second, we rightly want to avoid a "slippery slope" that would allow us to "explain away" any passage of Scripture.

In order to avoid the first pitfall, any argument for a new interpretation of Scripture should present a *positive* case; that is, it should not simply "explain away" apparently obvious meanings of Scripture. It should show thematic consistency with all of Scripture, a truly biblical worldview. To avoid the second pitfall, a new interpretation should delineate boundaries, defining what is negotiable and what is not. I hope to accomplish both of these tasks in the following pages.

My goal is to help the church avoid the same errors in the debate over the age of the earth that have occurred in the above examples. In the case of the moving earth, forcing an interpretation that the world does not move makes God into a great deceiver who shows us false appearances of things that are not real. In the same way, some people's views of the "apparent age" of the earth make God into a great deceiver. In the case of modern apostles, demanding that there are people who can perform the miraculous works of apostles can lead us to succumb to gullibility as we latch on to any claim that supports our position, and to fail to apply rigid tests to those who claim to have supporting evidence. In the same way, demanding that there must be scientific evidence for a young earth can lead us to latch on to people with dubious credentials who tell us what we want to hear.

In the case of the early return of Christ, if we insist on a rigid rule of the "most obvious" interpretation, we can cause people, including our children, to give up on the Bible, or reject Christianity outright as they lay what seems to be the most obvious interpretation alongside their experience. In the same way, every year the church loses children who go to college and find that the evidence does seem quite sound for an old earth, and who conclude they must reject the Bible. In the case of the Babylonian king, rejecting new information about biblical times means that we force a modern (and uninformed) view on the Bible instead of listening to those who are most familiar with the context of the ancient world. In the same way, reading Genesis 1 and other passages only in the way that seems most natural to modern eyes may cause us to lose some of the deeper meaning in those passages. In this book I will present some very deep themes of Scripture that often are lost in modern discussion.

We would do well to remember that science was founded by Christians who insisted that God is not a great deceiver, that the natural world is ordered by a good God, and that we must reject superstition and hearsay;

moreover, that we must subject all truth claims to rigorous examination, even claims of honored church leaders from generations past.[5] They insisted that the general revelation of God in nature and the special revelation in Scripture are in agreement, not discord. It is no coincidence that the scientific revolution and the Reformation came at the same place and time in history—the Protestants supported Kepler and Copernicus in their revolutionary new interpretation of the Bible. One could almost say that the Copernican revolution was primarily a revolution of Bible interpretation: it revealed that the scholars of the church past were not always correct in their interpretation of Bible passages like Psalm 93:1 (which had been interpreted to mean that the sun goes around the earth), just as they were not always correct in interpreting passages dealing with moral and spiritual issues.

As I said above, all scientific theories are provisional works of human beings, and therefore science does not "trump" Scripture, which is unchanging and inerrant. At the same time, all theological systems are provisional works of human beings, too. By "provisional," I do not mean "quickly changed." The scientific method provides rules by which theories may be changed, and successful theories last for centuries. In the same way, there are rules of Bible interpretation that do not allow us to easily jettison elements of theology. Yet the Reformation was based on the belief that the traditional teaching of the church about the meaning of Scripture is not to be confused with Scripture itself. Even the Roman Catholic Church, though it rejects the Reformation, now affirms that the understanding of the church can evolve and grow in the light of new information.

My view of "new" interpretations of Scripture in the church is the same as for individuals—we should be able to grow in wisdom, not rushing to every new wind of doctrine, but carefully weighing new views and always able to learn. A wise person finds new things constantly in Scripture, even while holding to it as an unshakable foundation, and the church does well to do the same. While we must not take lightly the Bible interpretation of faithful scholars of the past, we can also hope that new generations have something to add as well.

Review Questions

1. If you had lived in the year 1000, what would the most plain interpretation of Psalm 93:1 have been? How would you have felt if some scientist with a newly invented instrument called a "tele-scope" told you that the earth spun at thousands of miles per hour, and that this was compatible with the Bible?
2. If the most plain interpretation of Scripture, independent of our experience, is always the best, then is the most honest interpretation of Matthew 16:24–28 that Jesus would return to judge the world within forty years or so? Do you take a "possible" meaning of this passage because of your knowledge of things not reported by the Bible?
3. Is it legitimate to allow your experience with purported miracle workers to affect the way you interpret passages like Ephesians 4:11 and 2 Corinthians 12:12 that seem to promise signs and wonders?

The Scientific Case 2

My goal is to build a biblical case, not primarily a scientific one, but I want to first review some of the scientific facts so that we can see the stakes involved. As I said above, science does not trump the Bible, but the Bible has consequences for how we live in the real world, and one test of any theory of Bible interpretation is "Can we live with it?" Are we willing to accept the implications of our theology? We often apply this test: Can pacifists live consistently with their theology? Can those who believe that God promises health and wealth live consistently with this view? Some people have the impression that the implications for science in a young-earth view are relatively minor, for example, rejecting the validity of a few tests like Carbon-14 dating. I will argue that the implications are much more far ranging.

The Distance to the Stars

As a first example, let us consider the well-known measurements of the distances to the stars. These measurements work in "bootstrap" fashion. First, the measurement to near stars (out to about 3,500 light-years) are measured via "parallax." This is the same method you use when you move your head side to side to judge the distance to a far-off object. As the earth moves around the sun, the position of some stars in the sky changes in a way that depends on their distance.

Given this measure of distance to a set of stars, we can deduce a new distance yardstick based on the existence of a certain type of star called a "variable" star. These stars literally blink on and off, at a very regular rate. Careful measurements show that the rate of blinking is directly related to how much light they emit. Since the brightness of an object decreases with distance according to a well-defined law, if we measure the rate of blinking, and then compare the observed brightness to the amount of light emission that is implied by the blinking rate, this tells us the distance to the variable star. This method works for distances up to a few hundred thousand light-years. Variable stars are fascinating objects; it is almost as if God had put signal beacons in the universe as distance markers! Using these two distance measurements, we can deduce yet another yardstick via a statistical analysis that shows that stars at a certain distance have a well-defined "Hubble shift": the farther away they are, the more their color is shifted to the red. This yardstick can be used up to billions of light-years.

As most of us know, if light comes to us from an object that is a billion light-years away, then the light had to be traveling for a billion years. How do young-earth proponents explain these observations, then? There are three approaches:

- First, one could argue that the above measurement process is wrong, and that actually the stars are much nearer.

- Second, one could argue that the speed of light used to be much faster.
- Third, one could argue that the light we see did not actually come from stars, but was created "en route."

Each of these choices has important implications. If all the stars are clustered near us, no more than 6,000 light-years distant, then the heat and gravitational attraction they produced would have major effects on us. This is a point not typically well-appreciated by many young-earth creationists. Almost all people can easily accept the parallax yardstick out to 3,500 light years, but some people may have more trouble accepting the assumptions of the statistical analysis that gives larger distances. There are only relatively few stars within 3,500 light years distance, however. The vast majority of the billions of stars exhibit no parallax. If we suppose that these all hover around 6,000 light-years away, we must envision a vast halo of stars crowded upon each other just out of range of our distance measurement. Billions of stars all just 4,000–6,000 light-years away would have tremendous effects on us, and on each other as they crashed into each other. Every star attracts every other star due to the strong gravitational pull from their large masses. In the standard picture, the stars do not crash into each other because they are so far apart. If there were billions of stars all in a cloud just a few thousand light-years away, however, then they would necessarily have strong attraction to each other, crash on a regular basis, and emit huge radiation bursts when they crashed. Isaac Newton likened the stars to a collection of pins standing on end, connected by springs, and deduced that they must be immensely far apart to avoid hitting each other all the time.[1]

Of course, the crowding of billions of stars 6,000 light years away might not cause a problem if the laws of physics are dramatically different there. Then all bets are off—the radiation and gravitational attraction they produce could be whisked away by some unknown new laws of physics. Every analysis of the light from the stars indicates that the laws of physics are the same everywhere, however. For instance, the test of spectral analysis, which is used to give exact identification of chemicals, and which underlies much of the chemical and pharmaceutical industries, can identify elements in other stars that are exactly the same as those in our sun. In addition, the abundances of the elements observed in the stars agree to great accuracy with the predictions of nuclear theory, which underlies the entire nuclear industry.

Doubting the statistical measurements of distance does not completely let one off the hook in regard to the long distances. Although standard parallax measurements only work up to 3,500 light years, there is another geometrical method, similar to parallax, for measuring distance to the stars up to a few million light-years. This method works by looking at expanding objects, for example, the remnants of stars after they have exploded; it uses a standard method known as the Doppler effect to measure the speed of the expansion (the same method is used by police to measure the speed of cars, and by baseball parks to measure the speed of baseball pitches). Knowing the speed of the expansion, and assuming it is the same speed in all directions, one can measure the sideways expansion and thereby deduce the distance to the exploding object.[2] Distances measured this way are typically a few million light years. The only way to doubt this type of distance measurement is to assume that all these objects expand much more rapidly toward us than they do sideways—again, this amounts to assuming that God specifically rigs the universe in order to deceive us.

The second hypothesis above, that the speed of light used to be faster, would involve a change in all the laws of electricity and magnetism. Even a slight change in the speed of light over time would imply major changes in the calculations of electric and magnetic fields, calculations that underlie the entire electronics and communications industries. The speed of light is not independent of other physical effects. Many phenomena, including radio, light, magnetic and electric fields, X-rays, and friction, are all described by one set of elegant and simple equations, known as Maxwell's equations. These equations were created by James Maxwell, a Scottish Presbyterian believer with a strong faith in the beauty and simplicity of God's creation. The unification of all these effects into one set of equations means that one cannot propose a change in the speed of light without also proposing changes in everything from the structure of atoms to the color of the sun to the cost of computer circuits.

In fact, the hypothesis that the speed of light has changed substantially over time was suggested in the past century, most famously by Paul Dirac; this hypothesis was tested substantially and eventually rejected. Dirac wanted to believe this hypothesis because certain constants of nature, including the speed of light, appear to be "rigged" with very improbable values.[3] His hypothesis was ultimately rejected when it was found one cannot change the values of any of the constants of nature by more than a fraction of a percent without making life as we know it impossible. For example, the energy emitted by the sun depends sensitively on the speed of light; if the speed were a little faster, the sun would burn out rapidly.

This was actually the beginning of a very well-known line of argument for design in the universe. Our present science indicates that the constants of nature have very special values that cannot be changed without drastic effects; this seems to indicate design. The special values of the constants of nature form the basis of an apologetic argument known as the argument from design.[4] By contrast, those who say the speed of light has varied by a factor of a million or so since the beginning of the universe without any major problems for life on earth essentially are saying that the speed of light and other constants of nature can have any old value and are not special at all.

The third possibility, that God created the light "en route" to look as though it had come from stars that had exploded or collided but that never really did, raises the whole issue of "apparent age." This view is intellectually viable, in my opinion, and should be taken seriously. As many people have argued, if Adam and everything else had been created instantaneously, and Adam was created as a thirty-year-old adult, and not a baby, then the world would have looked at least thirty years old, because it contained at least one thing with that apparent age. If he had a belly button, the world would have looked even older, since it would look like he had a mother. Is not the creation of light en-route from the stars the same thing?

This argument is strictly irrefutable, just like the argument that all of us were created five minutes ago, including all our memories. In this case, however, one can hardly fault someone for working under the assumption that things are as they appear. I may in fact have been created five minutes ago, with all my memories, but who can blame me for operating under the assumption that my memories record real events?

The apparent-age argument obviously eliminates the possibility of any scientific discussion of the age of the world. Yet this is not typically the way young-earth proponents have argued. They typically argue that the universe actually *looks* young, only a few thousand years old, via evidences such as the thickness of dust on the moon being too small or the length of the Mississippi river delta being too short.[5] They would even say that the evidence for a young earth is so obvious that all modern scientists must effectively participate in a conspiracy to suppress the evidence. In so arguing, they implicitly admit that a scientific argument is a valid one, and they admit that agreement with our experience is important, as I have argued in chapter 1. One cannot pick and choose the data, however. If you want to present a scientific argument in one place, you cannot use an "apparent age" argument in another place. If you invoke apparent age to explain the light from the stars, then you must altogether stop trying to use scientific arguments to prove the earth is young. It is like the Wizard of Oz saying "look at my impressive evidences of credibility—but ignore that man behind the curtain!"

Many young-earth "scientific creationists" recognize this, and have stuck with option two, to try to present scientific arguments that the speed of light has changed over time.[6] In so doing, they are undertaking no small task. As mentioned above, to alter Maxwell's equations in order to get a change of the speed of light by a factor of a million within a thousand years or so would require a complete revision of all of modern science, including the laws of physics that underlie the communications, electronics, chemical, pharmaceutical, nuclear energy, and defense industries. Adjustments large enough to give the size of effect envisioned by the young-earth creationists would be easily observed, unless God chose to mask them perfectly, but this idea undermines the whole approach of the "creation-science" proponents, which is that the evidence for a young earth should be obvious.

In my opinion, of the above three choices, only the "apparent age" view is intellectually viable. On one hand, this view puts the proponents of this view in the same position as the Catholic Church at the time of Galileo, insisting on a sharp division between experience and the Bible, between "appearances" and "reality." This view also underlies one version of the Roman Catholic view of transubstantiation in the Mass—the "appearance" of the wine and bread stay the same, but their "essences" really change into the body and blood of Christ. Although this view is intellectually viable, the Protestant Church from the time of the Reformation has historically rejected such a view; the Catholic Church has also tended to limit this type of argument to true mysteries. As Francis Schaeffer has argued (for example, in his book *Escape from Reality*), such a sharp line between the physical world of our experience and the "essences" of the spirit can lead to a disjointed worldview in which the teachings of the Bible eventually become irrelevant, because they have nothing to do with the world of our experience, which most people think of as the "real" world.

On the other hand, some level of faith in "things unseen" is intrinsic to the Christian faith. In particular, we believe that the second coming of Christ could come at any time, even though the physical world looks like it will last for millions of years. Should we try to "prove" that Christ will return by analyzing societal trends for evidence of the Beast, estimating the amount of fuel left in the sun, or looking for new physical laws that will cause the stars to fall out of the sky?

One possible position, then, is to view both the creation and the second coming as off-limits to science, as fundamental points of faith that cannot be deduced from observation but are known only by the Word of God. This is not the position of the "scientific creationists." They argue that we can actually deduce that the creation happened recently by analyzing the scientific evidence properly. In my experience, many of these people also try to "prove" the second coming by various evidences such as the addition of a tenth nation to the Common Market in Europe. In my opinion, the quality of the standards used in these evidences is similar to the quality of the evidences used in "scientific creationism"—sensational and popular, but with a long record of retractions.

There is a difference between the second coming and creation in terms of experience, however. Time in the universe runs forward, not backward; therefore, we remember the past, but not the future. I can find physical evidence in my house of a robbery yesterday, but not of a robbery tomorrow. In the same way, the picture of faith in the Bible is not a blind leap to trust in an unknown god, but *remembering* God's faithfulness in the past as the foundation for our steps of faith into the unknown future. So there is reason to treat the physical past of the world as a subject of science, without giving science the same role in predicting the entire future.

Time Clocks in Creation

Further problems with the theory of apparent age come up when we look at other areas of science. One might not have trouble with the idea of light recording explosions of inanimate stars that never existed, but what about God burying the bones of animals that never existed? To go back to the example of Adam's apparent age, it is one thing to say that Adam might have appeared to be thirty years old the day after he was made, but what would Adam have said if he had thirty years of memories of events that had never occurred? Suppose that he had a memory of a mother, and thirty years of work and play, that he lived in a house that he remembered building, but God told him these things had never happened. Would he not have felt that God was playing tricks on him, that nothing he experienced could be trusted? The *level* of apparent deception makes a difference.

Imagine that you cut down a tree and count the number of rings. You know that the tree adds one ring each year, so you reason that the tree has existed a number of years equal to the number of rings.

Now imagine that the tree has one hundred thousand rings. Wouldn't it be reasonable to conclude that the tree had existed for one hundred thousand years? At the very least, you would have to say that the tree "appeared" to be one hundred thousand years old.

You will probably say, "There is no such tree!" But there are many things similar to tree rings in nature. For example, the Great Barrier Reef off the coast of Australia has millions of layers of coral. Knowing the rate at which these layers are added, one can use boreholes to calculate the age of the Reef; recent drills estimate the Reef is from six hundred thousand to a few million years old.[7]

Another example is the sediment layers deposited annually on the bottom of lakes. The following is an extended quote from *A New Look at an Old Earth*, by Don Stoner:

> Sediments which accumulate in the bottoms of lakes pile up in layers just like the growth rings of a tree. The Green River Formation of Utah, Colorado, and Wyoming is comprised of the sediments of a lake which dried up long ago. The formation is estimated to contain more than four million of these layers. I have spent some time counting them myself and can assure you that the published estimate is close enough to the actual truth for our purposes.
>
> If you don't trust my estimate, these layers can be viewed alongside U.S. Route 191, between Duchesne and Price in Utah. The formation extends southward from the

southern border of the Ashley National Forest. Because the highway drops steeply to the south, and because the layers of the formation tilt in the opposite direction, about a mile of cross-sectional thickness of the formation is exposed in just a few miles of road cut. Because of variations in climate over the time the lake existed, different layers can be quite different in appearance. Some are indistinct and difficult to count (a carefully aimed blow from a rock hammer fixes this), and a significant fraction of the layers (enough to prove the point all by themselves) are trivial to count (assuming a person has a great deal of time to spend); the individual layers are so sharply defined that they easily break apart into distinct flakes which are each about the thickness of a potato chip. The roadside is littered with these loose flakes which are broken off as the exposed formation weathers.

Geologists have many different theories about how long it took to deposit these layers. The standard theory is that each layer took a year to deposit—just like rings on a tree. Other geologists believe individual storms may have been responsible for many or most of the layers. Young-earth geologists point to extraordinary circumstances near construction sites etc. and argue that as many as three to five layers might have been deposited in a single day. But everyone has to admit that the layers are there—literally millions of them, neatly stacked in a huge vertical sequence—and that they were deposited at the bottom of a lake which once covered approximately 10,000 square miles.

Other suggestions have been made, such as the theory that these layers resulted from an erupting volcano like Mount Saint Helens. Although volcanoes can generate striped patterns in the ash they expel, it hardly needs mentioning that this ash is never comprised of calcium carbonate or neatly interlayered with fish fossils (as the Green River Formation layers are). Furthermore, the Green River Formation contains two thin layers which really are volcanic in origin. These two layers are distinguishable from the rest of the formation only because the rest of the formation is comprised of lake-bottom sediments rather than more of the same volcanic material. The details in the evidence don't allow for this sort of misinterpretation.

We will have to consider the possibility that these layers were all deposited during the single year of Noah's flood. In order to deposit four million layers in one year, the layers must be deposited faster than one every eight seconds! (The arithmetic always gives this same result.) The problem becomes one of changing the composition of the fine-grained sedimentary particles in an area about the size of Lake Erie from limestone to organic then back to limestone again, once every eight seconds. Furthermore, this must be done without disturbing the water so much that fine-grained sediments cannot settle out of it. This is clearly not what the evidence is telling us either. Remember, we are attempting to avoid miraculous false appearances here.

If we assume that the layers were deposited naturally over 10,000 years (ignoring Noah's flood and the fact that the lake vanished long ago), and if we assume the layers somehow formed faster than one per day, then we would still need to account for the mile depth of the formation. The Mississippi and Colorado rivers, taken together, drain about half of the area of the continental United States (excluding Alaska), and both are well known for the large amount of silt they carry (although before the Europeans arrived, they did not transport nearly so much silt). Together, even at today's rates, these two rivers transport less than a tenth of a cubic mile of sediments per year. If we made the impossible assumption that both of those rivers dumped their entire sediment load into this one lake and that no sediments ever washed out of it, this would only add about half an inch per year spread over the roughly 10,000-square-mile lake bottom. Even with these unreasonably extreme assumptions we cannot explain the evidence we see; in 10,000 years, this would account for less than a tenth of the formation's thickness.

And this is only a very small part of the bad news. All of the layers of the Grand Canyon can be shown to have been deposited before the first layer of the Green River Formation was. They extend right underneath that formation.

Please notice that I have made no "circular" assumptions about how long each layer was presumed to have taken to form etc. I gave the young-earth assumption every possible allowance. Still, a literal reading of this evidence simply will not allow for the young-earth theory of Biblical interpretation—with or without help from Noah's flood. Either God's creation testifies that it is much older than 10,000 years, or God has deceived us in his creation (fish fossils and all).[8]

Another powerful example from geology is the discovery of magnetic stripes on the floor of the Atlantic Ocean. This story is as follows: when the U.S. Navy started to engage in submarine warfare, Navy boats began to map the magnetic fields in the Atlantic Ocean in order to be able to tell the difference between the magnetic fields created by moving metal submarines and the background magnetic field. In the center of the Atlantic Ocean lies a high ridge on the ocean floor. The Navy mappers found, to their surprise, that every ten miles or so on each side of this ridge, the direction of the magnetic field background changes directions.

Scientists analyzing the magnetic orientations of the particles embedded in the rocks found that the orientation of these magnetic particles switches direction, and the locations of the switches in direction are symmetric on each side of the central ridge (see Figure 1). Why should this be? It turns out that the ridge at the center of the ocean is a point at which lava flows up to the earth's crust from the center of the earth. While this lava is still liquid, the magnetic particles are free to move and they line up with the earth's magnetic field. After the

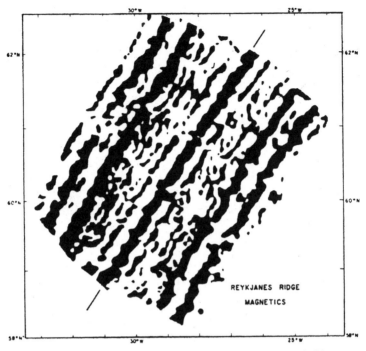

Figure 1. Magnetic domains on the Atlantic Ocean floor, recorded by Navy researchers in the 1960s who were using the magnetic information to track enemy submarines, not to prove anything about the age of the earth. The symmetry about the central ridge, indicated by the straight lines, is apparent even to a casual observer; statistical analysis has shown this symmetry rigorously. Reprinted from F.J. Vine, in *Plate Tectonics and Geomagnetic Reversals*, Allan Cox, ed. (San Francisco: W.H. Freeman, 1973).

lava cools and solidifies, they can no longer move, and their alignment is "frozen in." This solid lava then moves out from the center of the ridge as more lava comes up from below. This ultimately causes North America and Europe to drift further apart, the famous "continental drift" that is measurable with modern instruments at a few inches per year.

Why should the orientation of the magnetic particles change? Does the earth's magnetic field change directions? Independent research in the field of nonlinear dynamics has shown that the makeup of the earth and its magnetic core are such that the direction of the magnetic field does not stay the same, but does indeed change direction every million years or so, for the same reason that certain kinds of tops flip upside down as they spin. The spacing of the changes in orientation of the magnetic particles on the ocean floor corresponds to this amount of time, assuming that the continents have moved at the same rate that is measured today. The stripes of different magnetic orientation on the ocean floor therefore act as a "clock" with "ticks" every million years. Hundreds of these magnetic domains have been measured on the ocean floor.

What can we conclude, then? Either many scientists are guilty of a deception of incredible magnitude (including Navy mappers with no apparent axe to grind in the creation-evolution debate), or else the earth appears old according to a clock that has been verified with great accuracy. If the earth is not "really" old, then the level of apparent deception is disturbing. In the case of Adam's apparent age, looking about thirty years old the day after he was created, there is a reason for the appearances—Adam had to look like something. In the case of the magnetic fields on the bottom of the ocean, there is no reason why those have to be there just for us to exist.

Uniformitarianism vs. Flood Geology

Young-earth creationists often reject the assumption of "uniformitarianism"; in other words, the assumption that processes in geology have remained pretty much the same over the history of the earth. They argue that the assumption that one tree ring is added per year (or one inch of coral, or one layer on the bottom of a lake bed, or an inch of lava on the ocean floor) is an additional assumption that is not demanded by the data. They are correct in saying that this is an additional assumption, but they often ignore the problems of dropping this assumption. For example, to explain the magnetic domains on the ocean floor, some young-earth advocates postulate a short period of intense geological activity after Noah's flood, during which the continents moved several miles per year, with the earth's magnetic pole flipping from north to south every few years. The amounts of energy released in such motions are enormous, however. To imagine the energy entailed in this scenario, think of the energy of the last major earthquake in California, which moved that area a few centimeters, and then imagine how much energy would be required to make a whole continent plow through ten miles of solid rock per year, or twelve feet per day. The laws of nature would have to be utterly different in order to allow the preservation of life through such a phase in which energies greater than thousands of nuclear bombs were released. Of course, we can always suppose that God did a miracle to preserve life during this time, but there is no mention of either this intense continent-moving time or a miracle of preservation in the biblical texts. Instead of taking the simplest reading of Scripture, these young-earth advocates must read between the lines to get a whole new story.

Those who hold to flood geology may argue that the continents used to move faster, that lakes laid down sedimentary layers faster, and coral used to grow faster, but some fossil formations present problems for this view even if we allow for faster processes. For example, suppose that you came across the set of geological layers shown in Figure 2.

What could we make of this? Those holding to flood geology argue that all of the layers of fossils we find

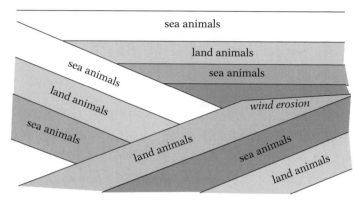

Figure 2. Hypothetical set of geological layers.

in the earth can be attributed to Noah's flood. Could a flood deposit the bones of animals in this way, sorted by land and sea? How did wind erosion occur under water? How could the layers be at different angles? Or were there two or more floods? Alternatively, one could argue that these layers were deposited in normal times, after the flood. How so, if we find these layers at the top of a high mountain? Should we propose that mountains have popped up from the earth at rates millions of times faster than seen today, similar to the way that light is supposed to have been millions of times faster at one time? If so, the energy released would be enormous, comparable to the amounts needed to move continents, discussed above.

Of course, one could propose that God did a miracle by sorting the fossils of the animals under the waters of the flood to precisely look as though they had died over millions of years. This is essentially the same as the apparent-age view that the bones belong to animals that never existed—God has worked a huge, miraculous deception to make the earth look old.

Figure 2 is just drawn from my imagination. Does

your view of the age of the earth demand that no such set of layers could ever occur in nature in reality? What would it mean for your view if you did find such a set of layers? Many geologists say that sets of layers equivalent to this do in fact occur in many places all over the world.[9] Young-earthers may dispute that, but the point of my imaginary example is that we cannot divorce our views from observation of the real world—if such a set of layers does exist, then the young-earth interpretation is extremely problematic.

In any case, the assumption of uniform behavior is hardly a wild hypothesis. We see lake beds, or coral layers, or magnetic domains on the sea bed, and all the layers look the same. There is no change in the appearance of the layers that makes some appear to have been deposited thousands of times faster than others. Without any additional information, it is entirely natural to assume that the layers were all laid down in the same way, just as we assume with tree rings.

It is, however, a mistake to say that scientists are committed to uniformitarianism at all costs. When evidence accumulates of dramatic past cataclysms, scientists have changed their views. In the past twenty years, many geologists have come to embrace the view of Luis Alvarez that a massive meteor struck the earth and killed most of the dinosaurs. In a classic example of the scientific method at work, Alvarez and others noticed a thin layer of rare iridium that occurs in the same geological layer around the world. Since iridium is not abundant on earth, they hypothesized that this layer was deposited by a meteor impact. This led to the prediction that a huge meteor crater should exist. The crater has since been discovered under Mexico's Yucatan Peninsula.

The topic of flood geology also raises several issues of Bible interpretation, which I will discuss in chapter 8.

Conspiracy Theories

Anyone reading the above scientific discussion must trust that the scientific facts I have presented are essentially correct. Some people may prefer to believe that thousands of non-Christian scientists are involved in a conspiracy to fabricate geological data. One thing acts as a strong check to prevent them from doing that, however: self-interest. Geology underlies the oil industry, and the oil industry is interested in finding oil with pinpoint accuracy, not in creating a vast religious deception.

Moreover, some Christians fault the old-earthers for violating the scientific method because they deal with things that lie in the past, and therefore beyond the realm of falsifiable predictions. This is incorrect. The theory of continental drift (and therefore of an old earth) is a highly successful, predictive theory, used by thousands of people who put millions of dollars at risk in order to predict where to find oil and coal. Just as capitalism tends to make people work toward productive goals out of self-interest, so it also tends to keep them scientifically honest, since a person who consistently denies reality and makes false predictions of where to drill for oil, at a cost of millions of dollars, will not last long in the business. If young-earth science made better predictions than old-earth science of where to find oil, I am convinced that the industry would embrace it in an instant.

Some people may believe that the scientific observations I have presented here are erroneous. These people must still answer the question, "What if these things were true?" Would it make a difference in your interpretation of the Bible? If you insist that proper science must give agreement with Scripture, so you hold out for evidence that the above facts are false, have you not conceded that failure to find such agreement weakens the case for a particular interpretation of Scripture? In that case you agree with my approach, and we are simply debating about the details. I only insist that if you admit this role for science, that you do not lightly overturn the conclusions of whole industries and academic disciplines without thinking through the implications.

I do not want to belabor the scientific case because many others have discussed these issues.[10] Unfortunately, much of the public and many theologians are simply not well equipped to make decisions about the scientific issues. I can only hope that when faced with various "scientific experts," Christians will apply as much skepticism toward those who say things that they want to hear as they do toward those who say things they don't want to hear. Truth is never served by gullibility or by picking and choosing evidence. The same tests should be applied to young-earth "creation scientists" as ought to be applied to self-proclaimed miracle workers or prophets—does the whole story check out, or are we just given anecdotes? Do we have to "pay no attention to the man behind the curtain"? People like Hal Lindsey, Henry Morris, and Michael Drosnin sell lots of books that seem to give support to Scripture, but do the facts check out? Do these books encourage faith or credulity? If hundreds of people of apparent integrity say that a person is in error, is it not possible that he is in error?

Capitulation to Evolution?

In this chapter, I have argued that the world looks as though animals and plants have been living and dying for millions of years. I want to stress, however, that this view is not synonymous with "evolution." On the issue of the origin of life, the positions of Darwinists and creationists are exactly reversed. Science is full of examples of "apparent design," both in the design of the physical world to allow life and in the construction of living creatures at all scales from the microscopic to the macroscopic.

Numerous books from the intelligent design movement (for example, those by Robert Shapiro,[11] Michael Behe,[12] Phillip Johnson,[13] and others[14]) present an impressive amount of evidence. I have studied biophysics and find it incredible that any honest person can deny the obvious design in every creature. Darwinists respond in a fashion similar to some young-earth creationists today, that the design we see is only "apparent." The dogma of Darwinism insists that we must have faith that future theories will explain everything in terms of random forces, despite all appearances.

Because these two issues are often conflated, many debaters end up talking past each other. The young-earth creationist presents many convincing evidences of design, making the evolutionist look foolish. Then the evolutionist presents many convincing evidences of an old earth, making the young-earth creationist look foolish. Neither one has addressed the other's argument. Evidence of design of living things is a *non sequitur* in regard to the age of the earth. The age of the earth is a logical *non sequitur* in regard to the appearance of design from random forces.

Many people seem to assume that if the earth is as old as science indicates, and animals have lived and died during that time, then evolution must have happened. Not so! If I placed a hundred high school students in a locked, empty room and told them, "I will shoot one of you every hour until you produce for me from scratch a complete copy of the works of Shakespeare and Handel," would I succeed in getting what I wanted? Certainly the desire for survival would provide a clear motivation for the students to try to please me, but if they did not have the ability to do what I wanted in the allotted time, they would all just die off. If I increased the time per student death to one hundred hours, would I have any better chance of getting what I wanted? In the same way, merely positing that animals have tried to survive over a long period proves nothing about the origin of their design. To prove evolution, one must show that *enough* time has passed for *known* random causes to produce the design actually observed. Modern science is far from doing this. Millions of years may seem like a long time, but life is incredibly well designed. How many millions of years would you need to randomly throw sticks of dynamite among paint cans near a canvas, before you reproduced the *Mona Lisa*?

I therefore have no qualms in saying that evolution by random forces alone is highly improbable from a scientific standpoint, even given millions of years of earth history. Some Christians have embraced "theistic evolution," the idea that God miraculously caused evolution to occur. Most Christians would agree that God could have created life this way, since he can do anything. I see no compelling scientific reason to take this position, however. As I have said, an old earth is not synonymous with evolution. The Bible texts also do not seem to support this position. The language of Genesis 1 seems to underscore the miraculous nature of creation, of God speaking and commanding the universe into existence, just as Jesus commanded the lame to get up and walk.

Saying that the earth is old does not mean that we do not see God's hand in the universe. On the contrary, we can give glory to God for the "age-old hills" (Gen. 49:26; Deut. 33:15 NIV) that demonstrate the majesty of the Ancient of Days, just as we give him glory for the beauty and design of living things.

Review Questions

1. Does it make any difference in the debate about transubstantiation whether the wine and bread actually appear to be flesh and blood? Is it legitimate to say that the elements are "really" flesh and blood but only "appear" to be bread and wine?
2. Should we apply tests of consistency and track record to people who claim to support the Bible, or should we present a common front to the world and ignore inconsistencies in such people? If popular advocates of Darwinism make oversimplifications and exaggerated claims, does this make such practices excusable for Christians?
3. Do you think the oil and coal industries would continue to adhere to a completely incorrect model of geology in order to propagate an atheistic worldview, even though it costs them millions of dollars? What would happen if one company started to use "creation science" instead, and it started to find oil much more consistently?

The Biblical Case I

Animal Death

In the beginning, God created the heavens and the earth.

The earth was without form and void, and darkness was over the face of the deep. And the Spirit of God was hovering over the face of the waters.

And God said, "Let there be light," and there was light. And God saw that the light was good. And God separated the light from the darkness. God called the light Day, and the darkness he called Night. And there was evening and there was morning, the first day.

Genesis 1:1–5

I have argued in the previous chapter that there are two tenable positions from a scientific standpoint—either the earth is really old, or it simply appears old. If I became convinced that the Bible teaches recent creation, I would adopt the view of apparent age, because scientifically I would feel it utterly dishonest to argue that the world actually looks young. I would hope that those who disagree with my Bible interpretation at least agree to this; otherwise they stand open to the charge of intellectual dishonesty.

At this point I turn to make a positive case for an old earth from Scripture. To do this, I will first tackle just one issue: whether animals died before Adam and Eve sinned. In my experience, this is the fundamental issue of Bible interpretation caught up in this debate. The interpretation of the words "day" and "evening and morning" used in Genesis 1 is of secondary importance. (I discuss the interpretation of these words in chapter 7.)

If the debate were only about the interpretation of the time-words in Genesis 1, there would probably not be such strenuous resistance to an old-earth view. The issue that seems to always give pause is the objection that animals could not have died before Adam and Eve sinned. Clearly, any theory of an old earth must involve animal death before Adam and Eve. The whole point of an old-earth view is to say that things are as they appear, and the earth is full of fossils and fossil matter such as coral and limestone.

I argue that the Bible teaches that animals did die before the fall of Adam and Eve. If one concedes this point, then many of the objections to an old earth go away.

My starting point is simply the following: what world does Genesis 1 describe? Is it our world or another world? If an ancient Hebrew had read this passage, what world would he have thought he was reading about?

It seems obvious to me that any normal reader would take this passage as referring to our world—that the emphasis of this passage is God's sovereignty over the real things that exist around us—not some other world.

In chapter 2 of Genesis, Moses goes out of his way to identify rivers and place names that were familiar to his readers, which existed in their world. But here is the rub: in our world, there are carnivorous animals. If the world described in Genesis 1 and 2 is our world, why not assume that it includes carnivorous animals?

The existence of carnivorous animals before Adam and Eve, of course, would mean that some animals died before Adam and Eve sinned. Many Christians believe that all death comes from the sin of Adam and Eve, and therefore they argue that God did not create any carnivorous animals in Genesis 1. They must go to other passages of Scripture to make their case. Several passages are used, which we will now examine at length.

The Curse

> The LORD God said to the serpent, "Because you have done this, cursed are you above all livestock and above all beasts of the field; on your belly you shall go, and dust you shall eat all the days of your life. I will put enmity between you and the woman, and between your offspring and her offspring; he shall bruise your head, and you shall bruise his heel."
>
> To the woman he said, "I will surely multiply your pain in childbearing; in pain you shall bring forth children. Your desire shall be for your husband, and he shall rule over you."
>
> And to Adam he said, "Because you have listened to the voice of your wife and have eaten of the tree of which I commanded you, 'You shall not eat of it,' cursed is the ground because of you; in pain you shall eat of it all the days of your life; thorns and thistles it shall bring forth for you; and you shall eat the plants of the field. By the sweat of your face you shall eat bread, till you return to the ground, for out of it you were taken; for you are dust, and to dust you shall return."

The man called his wife's name Eve, because she was the mother of all living.

And the LORD God made for Adam and for his wife garments of skins and clothed them. Then the LORD God said, "Behold, the man has become like one of us in knowing good and evil. Now, lest he reach out his hand and take also of the tree of life and eat, and live forever—" therefore the LORD God sent him out from the garden of Eden to work the ground from which he was taken. He drove out the man, and at the east of the garden of Eden he placed the cherubim and a flaming sword that turned every way to guard the way to the tree of life.

Genesis 3:14–24

Young-earth creationists identify the "curse" of Genesis 3:14–24 as a major change in creation, at which point carnivorous animals appeared. Notice what is *not* in this passage, however. There is no discussion of a complete re-creation of the universe, no discussion of new species of animals such as sharks and vultures, etc. The emphasis is that things will be *harder*, but not utterly different—Adam's work will require "sweat" and Eve's labor pains will be "increased." Nowhere does it say that new species of animals will appear or that the entire order of the physical world will change.

Some young-earth creationists argue that the change from eternal life for animals to the beginning of animal death is not major, that the species were just "twisted" a little. Such an argument only indicates that they have not carefully thought through the vast implications of a change from a world without death to a world with death. If there was no animal death, then either animals did not reproduce or the world had infinite surface area to contain the infinite multiplication of animals. Non-reproducing animals would be utterly different from those we have today, but there is no indication of this in the text—the animals are told to be fruitful and multiply.

A world with infinite surface area would also be utterly different from our finite earth.

Alternatively, one might argue that God planned for Adam and Eve to fall into sin, and therefore there was no problem simply because there was not enough time for overpopulation before Adam and Eve fell. There are major theological implications to this view, which says, in effect, that God made a world that was unstable unless sin entered in. Even so, fruit flies, bacteria, and some other species multiply so fast that, without death, the world would already have been overpopulated with them by the time Adam and Eve met. Some people might not want to classify insects like fruit flies as animal species exempted from death, but the rest of Scripture, for example, the Levitical dietary laws, classifies insects as "animals," so there is no textual reason to make them an exception.

Not only that, but in our world, many animal species have designs that make them suited perfectly and uniquely for killing and death. In a world without death, what would anteaters eat? What would sharks eat? Or vultures? An anteater that did not eat ants, or shark that did not eat fish, or a vulture that did not eat dead flesh would be utterly different from one of those species now. The change in nature of these animals to eat grass or other plants would require a total re-creation of them, as any biologist will testify. Yet such an utter change of all species is not mentioned in Genesis 3:14–24. All we read is that thistles will proliferate.

Some people might be tempted to read this passage as a "how the snake lost his legs" fable, and deduce from this that animal species were indeed changed dramatically. Almost all evangelical commentators agree, however, that the snake in this passage represents an incarnation of Satan, and the curse on him that the offspring (literally, "seed") of the woman would trample his head represents his defeat by the power of Christ. The story does not tell us that snakes started to eat other creatures or that they lost their legs; it tells us that the proud tempter will be humbled.

To argue that the changes in the physical world at the curse were indeed vast, some young-earth creationists point to the dramatic changes of the second coming and the new heaven and earth as evidenced by analogy. In Isaiah 11:6–8 we read that "the wolf will lie down with the lamb" and the lion will "eat grass." Actually, a study of this passage indicates that "lion" and "lamb" are used as symbols of human enemies who are reconciled; this passage probably does not talk about animal life in heaven at all. Nevertheless, there is no question that the new heaven and new earth will involve dramatic physical changes. Although many people debate their exact meaning, phrases such as "all the stars of the heavens will be dissolved" and "the sky will be rolled up like a scroll" (Isa. 34:4 NIV; Rev. 6:13–14) and "See, I make all things new" (Rev. 21:5) indicate dramatic and complete change in the physical order.

This brings us to an important interpretive debate. There are two different models of the creation, fall, and new creation, as illustrated below:

View I

world of Genesis 1–2	world of Revelation 21–22
our world (digression)	

View II

	world of Revelation 21–22
world of Genesis 1–2 our world	

In the first view, the original world of creation and the new heaven and new earth are essentially the same world—heaven restores the lost, perfect world—and our present world is utterly different from either. To support this view, advocates point to the imagery of the Garden of Eden used in Revelation 22:1–3, which includes the Tree of Life, a river, and the statement that there will no longer be any curse.

In the second view, the original world of creation and our world are essentially the same, and the new heaven and new earth of Revelation 21–22 is utterly different from these. This view is supported by several key differences between the world of Revelation 21–22 and the world of Genesis 1–3, as shown in Table 1. First, Revelation 21:1 (NIV) makes clear that "there was no longer any sea," while Genesis 1 emphasizes the creation of the sea. Also, Revelation 21:23 and 22:4–5 emphasize that there shall no longer be "night " or "the sun and the moon," while Genesis 1 emphasizes the existence of darkness, held back by the lights from the sun and moon, and the balance of morning with evening, that is, nightfall.

Revelation 21:1 says that the "first" heaven and earth pass away when the new heaven and new earth are created—not the "second." In other words, the heaven and earth of Genesis 1 (presented in Gen. 1:1) are lumped in together with our present heaven and earth, as a unity that will be destroyed when Christ comes again to make all things new. There is no mention in Scripture of a major physical change of the world at the fall. There are only two creation stories in the Bible—Genesis 1 and Revelation 21.

In the second view, the Garden of Eden of Genesis 2 is *typological* of heaven but not equal to it. "Typological" is a theological term that refers to the way God chose some physical things in the Old Testament to represent

Genesis 1–3	**Revelation 21–22**
And God said, "Let the waters under the heavens be gathered together into one place, and let the dry land appear." And it was so. God called the dry land Earth, and the waters that were gathered together he called Seas. (Gen. 1:9–10)	Then I saw a new heaven and a new earth, for the first heaven and the first earth had passed away, and the sea was no more. (Rev. 21:1)
And God made the two great lights—the greater light to rule the day and the lesser light to rule the night—and the stars. (Gen. 1:16)	And the city has no need of sun or moon to shine on it, for the glory of God gives it light, and its lamp is the Lamb. (Rev. 21:23)
And God saw that the light was good. And God separated the light from the darkness. God called the light Day, and the darkness he called Night. And there was evening and there was morning, the first day. (Gen. 1:4–5)	And its gates will never be shut by day—and there will be no night there. (Rev. 21:25)
And the rib that the Lord God had taken from the man he made into a woman and brought her to the man. Then the man said, "This at last is bone of my bones and flesh of my flesh; she shall be called Woman, because she was taken out of Man." Therefore a man shall leave his father and his mother and hold fast to his wife, and they shall become one flesh. (Gen. 2:22–24)	For in the resurrection they neither marry nor are given in marriage, but are like angels in heaven. (Matt. 22:30)

later, deeper spiritual realities. We say the earlier thing is a "type" of the later thing. For example, the temple in Jerusalem is typological of the true temple of God in heaven, according to Hebrews 8:1–5; in the same way, King David was a type of the Messiah, Jesus, and the snake on a pole was a type of the cross (Num. 21:8–9). A type is not just a "symbol," because the type has its own real story.

The Garden of Eden gives us a picture of heaven in the same way that the temple in Jerusalem gives us a picture of the holiness of God in his heavenly throne room. Just as heaven is a place of God's presence, the Garden was a space of special protection made for human beings, where God walked with man. Outside the Garden lay

Genesis 1–3	Revelation 21–22
Now the serpent was more crafty than any other beast of the field that the Lord God had made. He said to the woman, "Did God actually say, 'You shall not eat of any tree in the garden'?" (Gen. 3:1)	Then I saw an angel coming down from heaven, holding in his hand the key to the bottomless pit and a great chain. And he seized the dragon, that ancient serpent, who is the devil and Satan, and bound him for a thousand years. . . . And when the thousand years are ended, Satan will be released from his prison and will come out to deceive the nations . . . but fire came down from heaven and consumed them, and the devil who had deceived them was thrown into the lake of fire and sulfur where the beast and the false prophet were, and they will be tormented day and night forever and ever. (Rev. 20:1–2; 7–10)
	But nothing unclean will ever enter it, nor anyone who does what is detestable or false, but only those who are written in the Lamb's book of life. (Rev. 21:27)

Table 1. Differences between the worlds of Genesis 1–3 and Revelation 21–22

the dangerous natural world, just as heaven is bordered by the "outer darkness" of cursing described by Jesus (Matt. 8:12; 22:13; 25:30).

The whole image of the word "garden" is of a separated, special place. Just to be clear, I want to emphasize that I believe that the Garden was a real, physical place. Like the Temple in Jerusalem, however, it was a concrete shadow-picture of a heavenly reality, not the same thing. The letter to the Hebrews teaches that it is proper to read many Old Testament things as typological of future things (e.g., Heb. 8:1–5).

There is no doubt that the language of utter, revolutionary change is used to describe the new heaven and new earth in Revelation 21–22, Isaiah 34, and other passages. It is precisely this language that is lacking in the description of the fall in Genesis 3:14–24. The curse language in Genesis 3:24 gives the picture of *outward* motion. As R. C. Sproul eloquently discusses in *The Cross of Christ* video lecture series, the theme of exile outward, away from the presence of God, is a constant picture of the curse in Scripture, culminating in the death of Jesus outside the city gates of Jerusalem. The language of Genesis 3:24 gives the picture of Adam and Eve entering a pre-existing "outer darkness," not a newly created world with utterly new species and physical laws.

How could this "outer darkness" exist in a world that God had proclaimed "very good" in Genesis 1? This brings us to the entire issue of the existence of evil and the definition of what is "good." If we want to be faithful to the Bible, we cannot just say that evil is anything we don't like. For example, the Bible says very clearly that God's judgments and the existence of hell are good things, not bad, bringing glory to God; moreover, it says that in heaven no one will be embarrassed by the existence of hell, but instead will give praise and thanks to God because of it (Isa. 66:22–24, Rev. 19:1–3). We might feel that God should not have created hell, that it is "bad," but God tells us otherwise.

In my view, the powerful forces that existed outside the Garden, which include darkness, the sea, and carnivorous animals, existed prior to the fall as judgments held in readiness, as visible threats to Adam and Eve of the contrast between their protected state of grace and the possible consequences of leaving God's presence. These powerful forces are "good" in at least two ways: first, like any threat by God under the law, they show forth his power and justice, and second, as is made clear by Genesis 1, these powerful forces are not chaotic and out of control, but are clearly under the control of God who placed them in a careful and beautiful balance. This is a constant theme in the Bible, which we will explore later. The forces of judgment are held in check and balance: the day balances the night, summer balances winter, the land stops the sea, the carnivorous animal is itself eaten.

Some people will object that hell is good because it is judgment on evil people, while forces of judgment held in readiness in the creation before the fall could not be good because humans had not sinned yet. In this regard, we must not forget that human beings were in a "probational" state, sometimes given the theological name of the "covenant of law." The dangerous forces in creation were a drawn sword of judgment, so to speak, displayed to Adam just as he was told, "In the day you eat of it you shall die" (Gen. 2:17). One might even ask how Adam knew what God meant when he said "you shall die." If everything was idyllic and nothing died, how would Adam know what death was? Everywhere else in Scripture, the threats of God for disobedience invoke very concrete examples; Deuteronomy 28:15–68 gives a long list of very visual images of curses, frustrations, diseases, and climate changes.

Even if Adam did not observe animal death in the Garden, one cannot object that powerful threatening forces could not have existed in his world, because God himself gives Adam a powerful threat—"In the day you eat of it you shall die." This is a fundamental difference between the Garden and heaven: in heaven everyone lives in a state of grace and there are no threats. Romans 1:20 supports this view: "For since the creation of the world God's invisible qualities—his eternal power and divine nature—have been clearly seen, being understood from what has been made, so that men are without excuse" (NIV). This verse says these things have been seen "since the creation of the world," not "since the fall of human beings." From the very beginning, the demonstration of God's power in nature is associated with the threat of the law of God.

Some people might accept that things like darkness and the sea might be visible dangers held as threats to Adam and Eve, but reject the idea that carnivorous animals could belong to that category, because carnivorous animals hurt other animals, not humans alone. How could it be just to have animals die when they had not sinned? Yet the animals still have not sinned even now. Their death before the fall could be no more or less just than their death after the fall. In either case, they die because of forces beyond their control, as a demonstration of one aspect or another of God's nature—his power, or his wrath.

Further evidence that the Garden of Eden was not the same as heaven is found in the status of human beings relative to the angels. Psalm 8:4–6, quoted in Hebrews 2, talks of the temporary state of humans as "for a little while lower than the angels" (Heb. 2:7). Eventually, however, according to the apostle Paul, we will "judge angels" (1 Cor. 6:3) and rule over them in glory. Did humans rule over angels in the Garden of Eden? No—they are given authority to rule over the animals only (Gen. 1:28). In the Garden, a demon (fallen angel) roams on the loose, tempting Adam and Eve; in heaven there are no demons, since they have all been cast into the lake of fire.

There are two points of biblical interpretation, then, that I think can be made very strongly. First, the language of the curse of Genesis 3:14–24 does not indicate a complete change of the physical world, but rather an exile into a pre-existing outer darkness. It may be that substantial physical changes occurred at the fall, but if so, the argument must be made from other Scriptures, because no such indication appears in Genesis 3:14–24. All the Genesis passage says is that there will be harder work and more weeds, hardly a sweeping description of a new creation. Second, Scripture makes clear that the world of Genesis 1–2 is more similar to our present world than to the world of heaven. Humankind did not lose heaven at the fall—they were never in it. Instead, humankind was in a probational place of testing, a special garden of God's blessing surrounded by the forces of judgment, consistent with the probational state, or covenant of law, dictated to them by God.

This may seem strange to us because we have an idyllic view of the world of Genesis 1 and 2. I think that a Hebrew reader, however, would have instantly taken cues from several verses that dangerous forces existed from the start. Three things appear in Genesis 1 that almost always indicate danger in the rest of Scripture: the sea, the darkness, and the leviathan (the "great sea monsters" of Gen. 1:21 NASB). A person might argue that the sea, the darkness, and the leviathan in Genesis 1 were special "non-threatening" varieties that changed into new forms at the fall. But would not any Hebrew reading this passage have naturally taken those words to refer to the darkness, the sea, and the leviathan he knew?

We will return later to discuss the leviathan. Let us focus for a moment just on the symbolism in Scripture of the sea. Many people may be surprised to hear of the sea as a symbol of danger because they have fun memories of frolics at the beach. For the ancient Hebrew, however, the sea was a place of danger. Just as in the darkness, where dangerous animals lurk out of sight, ready to jump out, in the sea dangerous monsters lurk out of sight below the surface, ready to jump up. Also, the sea is vast and powerful, seemingly unending, just as the night sky is. The statement in Revelation 21:1 (NIV) that in heaven "there will no longer be any sea" makes sense in this context—there will be no darkness, and no sea. Yet both play a major thematic role in Genesis 1.

Table 2 lists examples of the sea used as a symbol for a dark force.

Exodus 15:19	For when the horses of Pharaoh with his chariots and his horsemen went into the sea, the Lord brought back the waters of the sea upon them, but the people of Israel walked on dry ground in the midst of the sea.
Job 7:12	Am I the sea, or a sea monster, that you set a guard over me?
Psalm 78:53	He led them in safety, so that they were not afraid, but the sea overwhelmed their enemies.
Psalm 93:4	Mightier than the thunders of many waters, mightier than the waves of the sea, the Lord on high is mighty!
Psalm 107:23–29	Some went down to the sea in ships, doing business on the great waters; they saw the deeds of the Lord, his wondrous works in the deep. For he commanded and raised the stormy wind, which lifted up the waves of the sea. They mounted up to heaven; they went down to the depths; their courage melted away in their evil plight; they reeled and staggered like drunken men and were at their wits' end. Then they cried to the Lord in their trouble, and he delivered them from their distress. He made the storm be still, and the waves of the sea were hushed.
Isaiah 5:30	They will growl over it on that day, like the growling of the sea. And if one looks to the land, behold, darkness and distress; and the light is darkened by its clouds.
Isaiah 17:12	Ah, the thunder of many peoples; they thunder like the thundering of the sea! Ah, the roar of nations; they roar like the roaring of mighty waters!
Isaiah 57:20	But the wicked are like the tossing sea; for it cannot be quiet, and its waters toss up mire and dirt.
Jeremiah 5:22	Do you not fear me? declares the Lord; Do you not tremble before me? I placed the sand as the boundary for the sea, a perpetual barrier that it cannot pass; though the waves toss, they cannot prevail; though they roar, they cannot pass over it.
Jeremiah 50:42	They lay hold of bow and spear; they are cruel and have no mercy. The sound of them is like the roaring of the sea; they ride on horses, arrayed as a man for battle against you, O daughter of Babylon!
Jeremiah 51:42	The sea has come up on Babylon; she is covered with its tumultuous waves.

Ezekiel 26:3	Therefore thus says the Lord God: Behold, I am against you, O Tyre, and will bring up many nations against you, as the sea brings up its waves.
Ezekiel 27:32	In their wailing they raise a lamentation for you and lament over you: "Who is like Tyre, like one destroyed in the midst of the sea?"
Daniel 7:2–3	Daniel declared, "I saw in my vision by night, and behold, the four winds of heaven were stirring up the great sea. And four great beasts came up out of the sea, different from one another."
Amos 9:3	If they hide themselves on the top of Carmel, from there I will search them out and take them; and if they hide from my sight at the bottom of the sea, there I will command the serpent, and it shall bite them.
Jonah 1:15	So they picked up Jonah and hurled him into the sea, and the sea ceased from its raging.
Zechariah 10:11	He shall pass through the sea of troubles and strike down the waves of the sea, and all the depths of the Nile shall be dried up. The pride of Assyria shall be laid low, and the scepter of Egypt shall depart.
Luke 21:25	And there will be signs in sun and moon and stars, and on the earth distress of nations in perplexity because of the roaring of the sea and the waves.
Acts 28:4	When the native people saw the creature hanging from his hand, they said to one another, "No doubt this man is a murderer. Though he has escaped from the sea, Justice has not allowed him to live."
Jude 13	Wild waves of the sea, casting up the foam of their own shame; wandering stars, for whom the gloom of utter darkness has been reserved forever.
Revelation 13:1	And I saw a beast rising out of the sea, with ten horns and seven heads, with ten diadems on its horns and blasphemous names on its heads.

Table 2. Examples of the sea as a dark and powerful force.

Death Came into the World

> Therefore, just as sin came into the world through one man, and death through sin, and so death spread to all men because all sinned—for sin indeed was in the world

before the law was given, but sin is not counted where there is no law. Yet death reigned from Adam to Moses, even over those whose sinning was not like the transgression of Adam, who was a type of the one who was to come. But the free gift is not like the trespass. For if many died through one man's trespass, much more have the grace of God and the free gift by the grace of that one man Jesus Christ abounded for many.

Romans 5:12–15

The above passage is often used to argue that all animal death started only after the fall—after all, doesn't Romans 5:12 say "sin came into the world through one man, and death through sin"?

This passage clearly teaches that Adam was one, real, historical man, not a symbol. But what kind of death came into the world when he sinned? Romans 5:12 goes on to say, "death spread to all men because all sinned." This passage clearly refers primarily to the death of humans, not animals.

Let us look closely at the story of how death came into the world. In Genesis 2:17, God says, "In the day that you eat of it you shall surely die." Did Adam die physically on that day? No, he lived another nine hundred years or so (unless one is willing to equate that "day" with an "age"). Then did God lie, and prove Satan correct when he said, "You will not die"? No—on that very day Adam died spiritually. Adam lost his soul. This was the primary death; physical death was a later consequence.

The concept of "spiritual death" is common in Scripture. The Bible often contrasts the dead heart with the living heart (Ezek. 11:19, 36:26). Jesus talks of people as "dead" though they are physically alive (Matt. 8:22), as does Paul (Eph. 2:1).

This kind of death is unique to humans. Only humans have the image of God (Gen. 1:27). The primary meaning of the death that came into the world when Adam sinned is therefore the spiritual death of alienation from God.

This spiritual death had the consequence of physical death—for Adam and Eve. When Adam and Eve were exiled from the Garden, they no longer had access to the Tree of Life. God specifically mentions this as a reason for exiling them (Gen. 3:22). Without the Tree of Life, they could not live forever.

Notice the significance of this: Adam and Eve had to eat from a special tree in order to have eternal life. Without the tree, they could not live forever. Did the animals eat from the Tree of Life? As far as we know there was only one tree, and it was located in the Garden (or Adam and Eve could have found one somewhere else), so animals outside the Garden clearly could not eat of the Tree of Life.

If physical death was impossible to any animal life before the sin of Adam and Eve, then why did Adam and Eve need to eat from a special Tree to have eternal life? Would not all animal life, including Adam and Eve, have already been immortal? The discussion of the Tree of Life seems to me to clearly indicate that eternal physical life was a special blessing given only to humans, not to animals. This is further supported by passages like Ecclesiastes 3:18–20 and Psalm 49:12, 20 that state that part of the curse on unbelievers is that "he is like the beasts that perish." In other words, animals naturally die, but humans should be different.

Above, I argued that perhaps animal death was an object lesson for Adam so that he would have some idea of what God meant when he said "in that day you shall die." If I am now arguing that the death Adam experienced on that day was spiritual, do I negate that argument? No—unless there were examples of spiritual death around for Adam to see. The physical death of the animals could serve as a visible token of the spiritual reality, as well as a concrete reminder of the physical death Adam would eventually suffer as a consequence of the spiritual death. The distinction between spiritual and physical death need not have been sharp in Adam's mind. Just as hell may be worse than fire, so the spiritual death Adam died was worse than the physical one, and using the latter as an illustration of the former can be quite proper.

Jesus undoes the work of Adam by bringing life where Adam brought death. How does Jesus do this? Does he give eternal life to all the animals? There is nothing in Scripture to warrant that belief, much as we might like to say that "all dogs go to heaven." No—Jesus brings spiritual life to humans right away, as Adam brought spiritual death right away, and Jesus brings physical immortality for humans as a consequence, just as Adam lost physical immortality for humans as a consequence.

Just as the symmetry of Romans 5:12–19 has been used to argue that just as Jesus was one historical man, so Adam was one historical man, one can also use the same symmetry to argue that Adam did not bring animal death unless Jesus brought animal life. Did Jesus die for every dog, ant, and bacterium? No—he shared in our humanity to die for "all men" (Rom. 5:18), not the animals.

Vegetarianism

> And God blessed Noah and his sons and said to them, "Be fruitful and multiply and fill the earth. The fear of you and the dread of you shall be upon every beast of the earth and upon every bird of the heavens, upon everything that creeps on the ground and all the fish of the sea. Into your hand they are delivered. Every moving thing that lives shall be food for you. And as I gave you

the green plants, I give you everything. But you shall not eat flesh with its life, that is, its blood.

Genesis 9:1–4

Many people have used Genesis 9:3 to argue that animals did not die before the fall. The argument goes as follows: in Genesis 9:3, God says to Noah and his family that they may eat animals just as they may eat of the plants. This verse is parallel in form to Genesis 1:29, in which God says to man that he may eat of the green plants. In the next verse, Genesis 1:30, God says to the animals that they are also given the green plants. The implication is that *only* the green plants were allowable as food before the flood because God makes a point of giving Noah the animals to eat in Genesis 9:3. Therefore, if no animals were eaten, no animals died.

This is an example of a "highly leveraged" argument. Several steps of logic are assumed, the failure of any one of which negates the argument. For example, the above argument seems to imply that no animals were eaten before Noah. Yet, animal death clearly preceded the flood. God killed an animal to make clothes in Genesis 3:21, and Abel's sacrifice of an animal is accepted by God in Genesis 4:4. The fact that Abel raised sheep also seems to indicate that he ate them, since this would be typical behavior for a Hebrew shepherd, and there is no indication in the text that Abel was anything other than a typical shepherd. Also, since many hundreds of years passed between Adam and Noah, it is fair to say that the world would have quickly overpopulated with animals if they had not died, unless the physical world was utterly different before Noah.

Many young-earth creationists place the change from vegetarianism to meat-eating at the fall of Adam and Eve. In this case, the statement in Genesis 9:3 is taken as an affirmation of an existing practice. But Scripture does not clearly state when the practice began, so identifying the beginning of this practice with the fall is only a conjecture. It could have been earlier. Clearly, there was no *sacrifice* of animals before then, but in many cases in the Bible God takes pre-existing things and gives them new meaning as covenantal signs: circumcision existed in Egypt before Abraham, people cooked and ate lambs before the Passover, criminals died on crosses before Jesus, people washed themselves before John the Baptist, and people ate bread and wine before the Last Supper.

Proponents of the young-earth view, however, would say that the statements of Genesis 1:29–30 rule out meat-eating before the fall. In so saying, however, they are arguing from a positive command to a negative one. They take the statement, "I give you x for food," to mean, "You are forbidden to eat anything else." This does not necessarily follow. The parallelism of Genesis 1:28–30 and Genesis 9:1–3 may be taken to imply the opposite, that is, that Genesis 9:1–3 is simply a repetition of the same charge given in Genesis 1:28–30, and that the expansion that includes eating animals is the same as other expansions of parallel passages in Scripture, such as the different versions of the Ten Commandments in Exodus 20 and Deuteronomy 5 or the Beatitudes in Matthew 5:3–12 and Luke 6:20–26. The later version does not negate the earlier version; rather, it is taken as simply saying the same thing with slightly more information. Note that the word "now" used in the New International Version translation of Genesis 9:3 does not appear in the Hebrew, nor does the past perfect tense; the English Standard Version, used above, is much more accurate. The NIV translation of this verse biases the reader toward the pre-fall vegetarian view.

If eating animals was not forbidden, then why would God single out green plants as food in the charge of Genesis 1:28–30? There are several plausible reasons. The best reason, in my opinion, is that God was establishing the hierarchy of his creation: green plants are the basis of the food chain. The sequence of life in Genesis 1 is

- green plants
- animals
- people;

in other words, from lesser to greater. In Genesis 1:28–30, God reverses this order. To human beings, he says, "Rule over the animals, birds, and fish, and rule over the green plants." To the animals, he says, "Rule over the green plants." It is similar to a military hierarchy: to the sergeants the general says, "Rule over the corporals and privates," and to the corporals he says, "Rule over the privates." The command to "subdue" and "rule over" the animals may plausibly be taken to mean "have power to kill"—that is what it generally meant among people (for example, Num. 32:29; Josh. 18:1; and 2 Sam. 8:11). And how else shall human beings normally rule over the birds and fish? Would an ancient Hebrew have understood this to mean mere obedience training?

Another plausible view is that these verses refer specially to life in the Garden, which I have argued was a special place separate from the rest of creation. Genesis 1:26–31 is, of course, parallel to Genesis 2:4–25. The latter story puts emphasis on the fruit trees in the Garden. Human beings, and the animals in the Garden with them, may have been specially commanded to eat only from the green plants and trees, while animals in the outer regions lived a different life.

Neither of these views may be acceptable to some people, but the fact remains that the teaching that human beings and animals did not eat meat before the fall is nowhere explicitly stated in Scripture; it is a deduction based on one or two verses that have alternate interpretations. It is always a dangerous practice to establish an orthodoxy by which men can be denied the pastorate based on a tentative interpretation of one or two verses.

Before moving on, I must mention that another problem with this view is that it implies that vegetarianism is a higher spiritual state that is practiced by those near to God but not by sinners such as we. Some might deny this implication, but if we take the command to humans in Genesis 1:28 to "rule over" the animals and strip it of the right to kill animals, it seems impossible to escape the moral authority this gives to animal-rights advocates and others who reject our right to kill animals. The implication seems inevitable that eating meat is a "lower" state, a product of the fall, and those who refuse to eat meat are closer to the "paradise state" of Genesis 1. Such a view seems to fly directly against the teaching of Paul (Rom. 14:2) that those who eat only vegetables are the "weaker brother."

I sometimes wonder if the objection to animal death before the fall has more to do with anthropomorphism than with real Scriptural arguments. It bothers us to think of animals dying for millions of years with no people around—the poor animals! Yet what is the difference between that and animals that die now? Unless animals are resurrected to eternal life, the lives of animals that die now are just as forgotten and vain as those that might have died millions of years ago. The Bible simply teaches that animals, like the grass, are here today and gone tomorrow, and that our lives are much more valuable than theirs: remember the words of Jesus, "Look at the birds of the air: they neither sow nor reap nor gather into barns, and yet your heavenly Father feeds them. Are you not of more value than they?" (Matt. 6:26).

Perhaps we also project human behavior onto animals when we think that carnivorous animals are "evil," and therefore could not have been part of a good creation. However, *predatory animals do not sin*. They simply do what they were designed to do. As St. Augustine wrote,

> One might ask why the brute beasts inflict injury on each other, for there is no sin in them for which there could be a punishment, and they cannot acquire any virtue by such a trial. The answer, of course, is that one animal is the nourishment of another. To wish that it were otherwise would not be reasonable. For all creatures, as long as they exist, have their own measure, number, and order. Rightly considered, they are all praiseworthy and all the changes that occur in them, even when one passes into another, are governed by a hidden plan that rules the beauty of the world and regulates each according to its kind. Although this truth may be hidden from the foolish, it is dimly grasped by the good and is as clear as day to the perfect.[1]

Bondage to Decay

> For the creation waits with eager longing for the revealing of the sons of God. For the creation was subjected to futility, not willingly, but because of him who subjected it, in hope that the creation itself will be set free from its bondage to decay and obtain the freedom of the glory of the children of God. For we know that the whole creation has been groaning together in the pains of childbirth until now.
>
> Romans 8:19–22

This passage is perhaps the single most-used "proof text" for those who reject the idea that animals died before the fall. It teaches that the creation is in a temporary state that will end when humanity enters into glory in heaven, at the new heaven and new earth following the second coming of Christ. The attributes of this temporary state are "futility," "decay," and "groaning." The passage clearly states that these things are not failures of God, but rather are in accord with his perfect will.

The passage does not answer the question, "When did this state come about?" A natural answer might be, "At the sin of Adam," but the passage does not say this. The pains are associated with the "childbirth" process of humanity, which ends when they enter "glory." The answer to the question of when these pains started is therefore the same as the answer to the question, "When did the human race become an unborn child, in the symbolism used here?"

In the analogy of this passage, the "born" child is the human race in "glory." Therefore, the answer to the question of when humanity was "unborn" is "when the human race was not in glory." This returns us to the issue we discussed earlier—was humanity in "glory" when they were in the Garden of Eden? Was the Garden substantially the same as heaven, so that we lost "glory" when we fell, and will regain it in heaven?

I have argued already that the Garden was not the same as heaven, but only a pale reflection of it, and that humanity was not in "glory" then. "Glory" as used in the Bible is associated with things such as ruling over the angels, having mansions, shining like the sun, flying in the air, etc. None of these things appear in Genesis 2—Adam is a humble gardener who rules over animals and plants, not a triumphant king who rules over angels and demons and stars.

In my view, then, humanity has been "unborn" since the beginning. Had humans not sinned, they would have passed through this probational state into glory at some point, to rule over the angels after "a little while." Jesus models what Adam should have done: live a sinless life, spend a short time in humility and submission, then ascend to glory, where he remains (Heb. 2:9).

The "futility" of the creation consists of the cycles of nature seen in things like the cycle of day and night, winter and summer, the ebb and tide of the sea, predator and prey. This futility existed as a sword of potential judgment held in readiness, which unhappily for us is now unsheathed and in operation. The book of Ecclesiastes eloquently describes this futility, or "vanity," in terms of these cycles in its first chapter:

> The sun rises, and the sun goes down, and hastens to the place where it rises.
> The wind blows to the south and goes around to the north; around and around goes the wind, and on its circuits the wind returns.
> All streams run to the sea, but the sea is not full; to the place where the streams flow, there they flow again.
>
> Ecclesiastes 1:5–7

Where else do these cycles appear? In Genesis 1! As I will discuss later, themes of cycle and balance, of danger held in check by God's provident hand, abound in the Bible, including the sea held in check by the land, the night held in check by the day, and the carnivorous animal that itself is eaten. All three of these themes occur in Genesis 1.

Martin Luther, although he believed in a recent creation, did not take this passage to imply a re-creation of the universe at the fall. Instead, in his commentary on Romans 8:20, he states that the "vanity" of the created order comes entirely from the change in our attitude toward these things:

> For all that God made "was very good" (Genesis 1:31) and is good to this day, as the apostle says in 1 Timothy 4:4, "Every creature of God is good," and in Titus 1:15, "To the pure all things are pure." It therefore becomes vain, evil and noxious, etc., without its fault and from

the outside, namely, in this way: because man does not judge and evaluate it rightly and because he enjoys it in a wrong way.... It is to this vanity, therefore (i.e. to this wrong enjoyment), that the creature is subjected.[2]

Luther clearly affirms that even today all created things are "very good" in and of themselves, including carnivorous beasts. This is a far cry from the view that every dangerous animal and even the second law of thermodynamics (the law of nature that leads to decay) is a warped, "bad" version of an earlier goodness. As Luther notes, 1 Timothy 4:4 says that the pronouncement by God of all things as "good" is still true; never in Scripture does God revoke this pronouncement. Instead, created things are "wearying" (Eccles. 1:8), "dangerous" (2 Cor. 11:26), or "empty" (Job 15:31; Isa. 40:17) to humans.

St. Augustine took a similar view. Discussing the thistles and weeds of the curse, he wrote,

> We should not jump to the conclusion that it was only then that these plants came forth from the earth. For it could be that, in view of the many advantages found in different kinds of seeds, these plants had a place on earth without afflicting man in any way. But since they were growing in the fields in which man was now laboring in punishment for his sin, it is reasonable to suppose that they now became one of the means of punishing him.[3]

The root meaning of the word translated as "futility," or "vanity," is temporariness or emptiness. A constant theme of the Bible is that the things of this world are passing away, and the true and the permanent will be revealed only in heaven. My interpretation of Romans 8:19–24, in agreement with Luther and Augustine, is that the original creation had this kind of temporariness, or emptiness, because it was designed as a place of probation, not as the permanent home of humankind. We see this as evil because we try to grab hold of this world too tightly. We lose sight of our eternal destiny. We try to find our satisfaction in the things of this world, only to be disappointed, because they do not have the kind of permanence that can satisfy our souls.

What about the "groaning," however? This seems to indicate something is wrong, not just temporary. There *is* something wrong with creation. But what is wrong is not the fact that animals die or trees decay. What is wrong is that the king of creation, the race of Adam, has abdicated his role. All of creation has been affected by this failure, both in the failure of the human race to take care of the creation properly, and in the increase of pain that God has sent into the world as a punishment. We do not have to imagine an entire reshaping of the world, with new laws of physics and new, carnivorous species of animals, for creation to be groaning.

It is perhaps not going too far to say that the increase of the woman's labor pains in the curse of Genesis 3:16 is symbolic of the increase of the labor pains of the whole world. Had Adam and Eve not fallen, Eve would certainly have become pregnant and born children, in fulfillment of God's command to be fruitful and multiply. In the curse, God does not say that he will create something utterly new by making her have labor pains. Instead, he says that her labor pains will be "increased." In the same way, the futility of creation is "increased" by Adam and Eve's sin (including things such as pollution, overuse, wars, and mismanagement of land) but not created utterly new. The creation has been "expecting" since the beginning of creation, but because of Adam and Eve's sin, it now has "labor pains" just as the daughters of Eve do.

To get technical, Romans 8:22 says that the creation has been groaning "up to now." The most direct reading of this is that Paul means "for all times up to now," not "from the fall up to now." One might argue that the latter reading is implied, but the passage does not say this—Paul simply says "until now."

Saying that all the futility and decay of nature is a result of the fall perhaps goes too far. Decay is a result of the second law of thermodynamics, which states that disorder always increases, that is, that everything must run down eventually. Is the second law of thermodynamics in our universe due to sin? If so, then time is due to sin, because time runs forward and not backward because of the second law of thermodynamics. This is a deep point of physics not always appreciated by non-experts. We can only tell the passage of time with a "clock," a natural cycle. But we cannot tell whether a cycle is running forward or backward unless there is decay. Thus the abolishment of decay literally means the end of time. Yet the description of Genesis 1 is full of time-clock references, most notably in the sequence of days, and the use of the sun and moon to mark the times and the seasons.

God could, presumably, make a world in which time existed but not decay. But such a world would have to have utterly different laws of physics from our own world—and the world of Genesis 1 sounds very much like it is supposed to be our world, not another.

Review Questions

1. If all life was immortal before the fall, why did Adam and Eve have to eat of a special Tree of Life to be immortal, according to Genesis 3:22?
2. How did Adam and Eve know what "death" meant when God told them, "In that day you shall surely die"?
3. If the one act of disobedience of Adam led to death for all animals, did the one act of righteousness of Christ lead to life for all animals?

4. If thousands of new, carnivorous species were created after the fall, why doesn't the Bible mention it? Is such an interpolation any different from interpolating that the days of creation may have been longer than twenty-four hours?
5. If 1 Timothy 4:4 says, in effect, that every creature of God is good, on what basis can one argue that carnivores today are "bad" and could not have been included in the pronouncement of Genesis 1:31 that all creation was "very good"?

The Biblical Case II

The Balance Theme in Scripture

In the previous chapter, I argued that animals died before the fall by arguing against the use of certain passages that have been taken to imply animal immortality before the fall. In this chapter, I want to build a more positive case for animal death before the fall by looking at other passages of Scripture that indicate that the physical world of Genesis 1 is substantially the same as the world we live in now, with the exception of the Garden of Eden.

I have already mentioned several times the balance themes of Scripture, which include darkness and light, sea and land, etc. Some might accuse me of promoting a "yin and yang" philosophy, but the Bible's idea of balance is quite different from the Eastern concept of yin and yang. Yin and yang and other similar pagan ideas start with an observation of something very obvious about nature—the beautiful balances that include day and night, land and sea, summer and winter, predator and prey, male and female, etc. Two pagan ideas are added to this: first, that these powers are beyond the control of God (if there is a God) and they are the effect of an irresolvable, chaotic warfare between irreconcilable opposites; and second, that good and evil acts by humans are part of this natural balance, so that even sin has a proper place in the cycle of life.

The Bible attacks both of these ideas, starting in Genesis 1: God stands sovereign over all the powers of nature, and evil acts by people have no proper place but are judged by him. Even so, the Bible does not deny the existence of natural balances. Instead, it praises God for the balance of nature, even of those powerful elements that can endanger us. Who among us has not been impressed with the balance of nature, in which decaying animals become food for plants, what is exhaled by plants is inhaled by animals, and water from the oceans returns to the forests on the mountaintops?

If any of my opponents deny the obvious balance-of-nature theme in Scripture, let them first turn to Genesis 8:22 (NIV), a poem entirely about balance:

> As long as the earth endures,
>   seedtime and harvest,
>   cold and heat,
>   summer and winter,
>   day and night
> will never cease.

The context of this poem is God's promise after the flood to never again destroy the land. This emphasizes an important aspect of the balance theme, that forces of judgment (in this case, the sea) are held in check by forces of blessing (in this case, the land). In this poem, several other balances are mentioned: the starvation of seedtime ameliorated by harvest, the cold ameliorated by the heat, the night ameliorated by the day.

Is this a product of sin? Let's turn back to Genesis 1 and look for the following balances: light and darkness, day and night ("evening and morning"), land and sea, male and female, seed and harvest.

> And God separated
>   the light from the darkness.
>   God called the light Day, and the darkness he
>   called Night.
>   And there was evening and there was morning,
> the first day.
>
> Genesis 1:4–5

> And God made the expanse and separated
>   the waters that were under the expanse from
>   the waters that were above the expanse.
> And it was so.
>
> Genesis 1:7

> And God said, "Let the waters under the heavens
>   be gathered together into one place, and let
>   the dry land appear." And it was so.
>   God called the dry land Earth,
>   and the waters that were gathered together he
>   called Seas.
> And God saw that it was good.
>
> Genesis 1:9–10

> And God made the two great lights—
>   the greater light to rule the day and the lesser
>   light to rule the night—
> and the stars.

> And God set them in the expanse of the heavens
> to give light on the earth,
> to rule over the day and over the night
> and to separate the light from the darkness.
> And God saw that it was good.
>
> Genesis 1:16–18

> So God created man in his own image, in the
> image of God he created him;
> male and female he created them.
>
> Genesis 1:27

This balance theme occurs almost everywhere in Scripture where the creation is discussed. Here I will discuss some of the most important passages.

Dangerous Forces in Creation

> Where were you when I laid the foundation of the earth? Tell me, if you have understanding. Who determined its measurements—surely you know! Or who stretched the line upon it? On what were its bases sunk, or who laid its cornerstone, when the morning stars sang together and all the sons of God shouted for joy?
> Or who shut in the sea with doors when it burst out from the womb, when I made clouds its garment and thick darkness its swaddling band, and prescribed limits for it and set bars and doors, and said, "Thus far shall you come, and no farther, and here shall your proud waves be stayed"?
> Have you commanded the morning since your days began, and caused the dawn to know its place, that it might take hold of the skirts of the earth, and the wicked be shaken out of it? It is changed like clay under the seal, and its features stand out like a garment. From the wicked their light is withheld, and their uplifted arm is broken.

> Have you entered into the springs of the sea, or walked in the recesses of the deep? Have the gates of death been revealed to you, or have you seen the gates of deep darkness?
>
> Have you comprehended the expanse of the earth? Declare, if you know all this. Where is the way to the dwelling of light, and where is the place of darkness, that you may take it to its territory and that you may discern the paths to its home? You know, for you were born then, and the number of your days is great!
>
> Have you entered the storehouses of the snow, or have you seen the storehouses of the hail, which I have reserved for the time of trouble, for the day of battle and war?
>
> Job 38:4–23

> Can you hunt the prey for the lion, or satisfy the appetite of the young lions, when they crouch in their dens or lie in wait in their thicket? Who provides for the raven its prey, when its young ones cry to God for help, and wander about for lack of food?
>
> Job 38:39–41

In Job 38–41, God gives an extended boast about his creation. The balance theme is strongly evident here, starting with the same balance that Genesis 1 does, the sea and the land (Job 38:8–11), in which God says to the sea, "This far you may come and no farther." The same language is used in Jeremiah 5:22 and in Proverbs 8:27–29:

> When he established the heavens, I was there; when he drew a circle on the face of the deep, when he made firm the skies above, when he established the fountains of the deep, when he assigned to the sea its limit, so that the waters might not transgress his command, when he marked out the foundations of the earth.

This is only one example of the theme in Scripture of dangerous forces that are "bounded" by checks and balances. Later in Job 38 we read of

- the bounding of darkness by light (38:19–20),
- snow and hail, which are kept in "storehouses" (38:22), and "reserved for the time of trouble" (38:23),
- drought and harvest (38:26–27),
- wind and storms (38:34–35) that send "lightning bolts" (cf. Jeremiah 10:13), controlled by the voice of God,
- predator and prey (38:39–41).

Some would argue that the dangerous forces mentioned in this passage, in particular predator and prey, are products of the fall. But one cannot miss the association of these things with the *creation*. Job 38 tells us the time period God has in mind: "when I laid the foundation of the earth" (38:4), when God "laid its cornerstone" (38:6). The chapter moves on from the creation to God's daily upholding of the world without a break, in one continuous discussion. There is no mention of these dangerous forces coming about due to the fall. Furthermore, things that are clearly present in Genesis 1 are specifically associated with judgment in Job 38—even the morning "takes the earth by the edges to shake the wicked out" (38:12–13 NIV), and the sea is identified with the "gates of death" (38:16–17).

One might argue that these things have a new function since sin has come, but in that case, why not also argue that predators have a new function? A predatory animal is no more intrinsically evil than a hailstorm or the sea—each can be dangerous to us, but each can also be a beautiful demonstration of God's power.

Continuing on in Job, chapters 39–41 mention several animals that God is proud of. These animals have numerous associations with death and suffering.

- The mountain goats and deer have "labor pains" (39:3 NIV),
- the donkey lives in a "wasteland" (39:6 NIV),
- the ostrich "deals cruelly with her young" (39:16),
- the horse strikes "terror with his proud snorting" (39:20),
- the hawks "suck up blood" (39:30),
- the leviathan has "fearsome teeth" (41:14 NIV).

Again, one might argue that these are all products of the fall, but how does God talk about them? Does he say, "Be ashamed of the evil you have brought into the world"? No, he brags about how they all glorify him; in other words, how they are "very good."

Was Leviathan Carnivorous?

The leviathan is particularly important here, because the leviathan appears in Genesis 1:21 as the "great sea creatures" (*tannin* in Hebrew, also translated as "sea monsters" NASB). There is no doubt that the great sea creature of Genesis 1:21 is to be equated with the leviathan, as an examination of Isaiah 27:1 and Psalm 74:13–14 shows:

> In that day the LORD with his hard and great and strong sword will punish Leviathan the fleeing serpent, Leviathan the twisting serpent, and he will slay the dragon [the *tannin*] that is in the sea.
>
> Isaiah 27:1

> You divided the sea by your might; you broke the heads of the sea monsters [the *tanninim*] on the waters. You crushed the heads of Leviathan; you gave him as food for the creatures of the wilderness.
>
> Psalm 74:13–14

The Old Testament often uses a literary device that wonderfully aids translation from the Hebrew: "parallelism." The same thing is said twice, in different words. Both of these passages use Hebrew parallelism to equate the leviathan with the *tannin*. The term *tannin*, used in Genesis 1:21, is variously translated "sea creature," "dragon," or "coiling serpent" and in general refers to a dangerous reptile. It is never used in a gentle, benign sense anywhere in Scripture.[1] Table 3 gives some additional examples of the use of the word *tannin*.

There is no doubt that the leviathan described in Job 41 is carnivorous and dangerous. Would a Hebrew reading Genesis 1:21 at the time of Moses have understood this term to refer to the leviathan he knew, the carnivorous, dangerous variety, or some unknown, grass-eating version? Would this term, "great reptile," have inspired in him feelings of awe at the dangerous powers created by God, or feelings of fondness for the peaceful ease and comfort God made? Is it not "reading in" to the text to take the sea monster of Genesis 1:21 as anything other than the carnivorous, dangerous version familiar to the Hebrews? Many people accuse old-earth advocates of "reading in" to the text what they want to see, but changing the sea monster of Genesis 1:21 into a docile, unthreatening animal seems to me an egregious case of overturning the plain meaning of Scripture.

The presence of the great reptile in Genesis 1 takes on even more meaning when one learns that other ancient cultures had creation stories in which great reptiles figured predominantly as agents of the formation of things. In Genesis 1 we see them mentioned in passing as another type of animal created by God. The message to an ancient Near Eastern reader is clear: God is in control of all those dark forces that you fear, including the sea and the great monsters.

Deuteronomy 32:33	Their wine is the poison of serpents [*tanninim*] and the cruel venom of asps.
Jeremiah 51:34	Nebuchadnezzar the king of Babylon has devoured me; he has crushed me; he has made me an empty vessel; he has swallowed me like a monster [*tannin*]; he has filled his stomach with my delicacies; he has rinsed me out.
Ezekiel 29:3	Speak, and say, Thus says the Lord God: "Behold, I am against you, Pharaoh king of Egypt, the great dragon [*tannin*] that lies in the midst of his streams, that says 'My Nile is my own; I made it for myself.'"
Job 7:12	Am I the sea, or a sea monster [*tannin*], that you set a guard over me?
Daniel 7:5	And behold, another beast [*tannin*], a second one, like a bear. It was raised up on one side. It had three ribs in its mouth between its teeth; and it was told, "Arise, devour much flesh."

Table 3. Uses of the word *tannin* in Hebrew Scripture.

As an aside, one can ask, "What was the leviathan?" What animal should we equate with this phrase? I have already mentioned several other words associated with it in Scripture: scales, sharp teeth, serpent, coils. The term is generic for "dangerous reptile." As such it is appropriate to equate it with the alligator or crocodile. But it is equally valid to equate it with the dinosaur, which seems to agree more with the description of Job 41. For example, in Job 41:34 (NIV), we read that the leviathan "looks down on" the rest of creation; the whole passage makes the leviathan seem huge. This doesn't mean that dinosaurs were walking around at the time of Job. It is not hard to imagine that people in ancient days could have discovered huge dinosaur skeletons, just as we have, and noticed the similarity to the great reptiles (e.g., the Nile crocodile, which can grow to fourteen feet) of their day. In fact, there may have been more dinosaur skeletons around in their day than now, if people removed the bones as curiosities or magic relics. The near-universal belief in dragons in different cultures indicates that there was some corporate knowledge of huge reptiles. The identification of the leviathan with dinosaurs is unimportant for our discussion, however; what is important is that the leviathan/sea monster as understood by the Hebrews was carnivorous and dangerous.

A common interpretation of Psalm 74:13–14 is to take the leviathan as a symbol of Egypt and not as a real animal. In my opinion, however, this is an example of over-allegorical commentary. While it is true that Egypt is symbolized by a serpent in many places in Scripture, nowhere in this Psalm does Egypt appear. Instead, it is another example of the "balance of creation" theme in Scripture:

- the leviathan, who eats, is eaten (74:14),
- rivers flow and then dry up (74:15),
- the day balances the night (74:16),
- God sets the "boundaries" of the land (74:17),
- God balances summer with winter (74:17).

The main exegetical reason for taking the leviathan as a symbol for Egypt comes from the mention of the division of the sea in verse 13, which is taken as a reference to the crossing of the Red Sea. The "division" of the sea also occurs in Genesis 1:6–7, however, as one of many important "divisions," (see also Prov. 3:19–20). The rest of the context of Psalm 74:13–17 indicates that the balance of the sea with the land is the topic as God opens and dries up the streams and rivers, and establishes the boundaries of the land. This is not a novel interpretation; John Gill (1697–1771), in his commentary on this passage, noted that several early commentators took the view that these verses referred to the division of the sea in the creation.

This psalm also contains another theme of Scripture, that of "wild beasts" as a judgment (Ps. 74:19; cf. Deut. 32:24; Jer. 12:9; Ezek. 5:17; 14:15, 21; and Rev. 6:8). Some would argue that carnivorous beasts could not exist in Genesis 1, since there were no sins to judge yet. But as discussed in the previous chapter, there is no second creation story in Scripture—Genesis 1 appears to list every type of animal being created, and then Genesis 2:1 says the creation was "finished, and all the host of them." The wild beasts existed as judgments held in readiness, just like the darkness and the sea, which are clearly used as judgments in the rest of Scripture (darkness: Exod. 10:21–29; Rev. 16:10; the sea: Exod. 14:26–15:8; Rev. 18:21).

If one wants to take seriously the literal or natural meaning of every word in Genesis 1, as young-earthers want to do with the word "day," then one must accept that the *tannin*, or sea monster, is a monster. There is simply no honest way to read this as referring to a gentle, non-carnivorous animal.

Praise for the Darkness

> Bless the Lord, O my soul! O Lord my God, you are very great! You are clothed with splendor and majesty, covering yourself with light as with a garment, stretching out the heavens like a tent.
>
> He lays the beams of his chambers on the waters; he makes the clouds his chariot; he rides on the wings of the wind; he makes his messengers winds, his ministers a flaming fire.

He set the earth on its foundations, so that it should never be moved. You covered it with the deep as with a garment; the waters stood above the mountains. At your rebuke they fled; at the sound of your thunder they took to flight. The mountains rose, the valleys sank down to the place that you appointed for them. You set a boundary that they may not pass, so that they might not again cover the earth.

You make springs gush forth in the valleys; they flow between the hills; they give drink to every beast of the field; the wild donkeys quench their thirst. Beside them the birds of the heavens dwell; they sing among the branches. From your lofty abode you water the mountains; the earth is satisfied with the fruit of your work.

You cause the grass to grow for the livestock and plants for man to cultivate, that he may bring forth food from the earth and wine to gladden the heart of man, oil to make his face shine and bread to strengthen man's heart. The trees of the LORD are watered abundantly, the cedars of Lebanon that he planted. In them the birds build their nests; the stork has her home in the fir trees. The high mountains are for the wild goats; the rocks are a refuge for the rock badgers.

He made the moon to mark the seasons; the sun knows its time for setting. You make darkness, and it is night, when all the beasts of the forest creep about. The young lions roar for their prey, seeking their food from God. When the sun rises, they steal away and lie down in their dens. Man goes out to his work and to his labor until the evening.

O LORD, how manifold are your works! In wisdom have you made them all; the earth is full of your creatures. Here is the sea, great and wide, which teems with creatures innumerable, living things both small and great. There go the ships, and Leviathan, which you formed to play in it.

These all look to you, to give them their food in due season. When you give it to them, they gather it up; when you open your hand, they are filled with good things.

When you hide your face, they are dismayed; when you take away their breath, they die and return to their dust. When you send forth your Spirit, they are created, and you renew the face of the ground.

May the glory of the Lord endure forever; may the Lord rejoice in his works, who looks on the earth and it trembles, who touches the mountains and they smoke!

Psalm 104:1–32

Psalm 104 is another important example of how God receives glory for his creation of forces of judgment. This entire psalm is strongly parallel in form with Genesis 1, starting with the creation of the light (104:2), and the heavens and the earth (104:2–5) with waters above and waters below (104:6–8), the seas that are given "a boundary that they may not pass," (104:9), and the beasts and the birds and human beings (104:11–14), then starting over as Genesis 1 does on the fourth day, with "the sun and the moon," the beasts, and human beings (104:19–23).

Some people have taken 104:6 as a description of the flood of Noah, when the "waters stood above the mountains," but it makes more sense in the context of creation. The phrase "the waters stood above the mountains" can easily be equated with the waters that stood above the expanse in Genesis 1:6.

Psalm 104 is a song of praise, and some of the things for which God is praised are the darkness (104:20) that releases the wild beasts and lions who seek for their prey (104:21) and the sea (104:25) that contains the leviathan (104:26). The balance theme is summed up in Psalm 104:27–30: "These all look to you, to give them their food in due season. When you give it to them, they gather it up; when you open your hand, they are filled with good things. When you hide your face, they are dismayed; when you take away their breath, they die and return to their dust. When you send forth your Spirit, they are created, and you renew the face of the ground."

Like Job 38, Psalm 104 is one continuous narrative. It does not give any indication of a sharp break between things that were created in Genesis 1 and new species or natural forces that supposedly appeared after the fall. The darkness, the sea, the leviathan, and the lion that catches its prey are all good things for which God is praised.

Most people have no trouble praising God for happy, pleasant things, but can we praise him for dangerous forces that are used for judgment? As discussed earlier, the Bible makes clear that God is not ashamed of his judgments, including hell, but is glorified by them.

Some people might argue, however, that things like the darkness, the sea, wild beasts, and the leviathan can be good as judgments in *response* to sin, but could not be good if held in reserve before humankind had sinned. But who are we to tell God what to do? I have argued that these things served as powerful reminders of the threats of the covenant of law, the probational state of Adam and Eve in which God offered a blessing or a curse depending on their obedience.

Someone might ask, "Then why should they exist for millions of years before humans?" The answer is that the universe dwarfs us in both age and extent, and this also teaches us of the majesty of God. There are millions of stars and planets that will be born and die in explosions so far away from us that we will never know of them, and there are wars fought by insects and microbes so small that we cannot see them. In the same way, millions of things may have passed in history without our knowing them. Even while humans are the highest member of God's creation, we must not lose sight of the fact that God is the center of creation, not us. The universe testifies of *God's* "eternal power and divine nature" (Rom.

1:20)—what better way to testify of this than a world so vast in both time and space that we can scarcely conceive of it?

Perhaps a person might say, "Well, if I wanted to make a universe that was 'very good,' I wouldn't include millions of years of animals suffering in it!" But who are we to tell God what to do? Isaiah 45:5–12 is especially important in this regard:

> "I am the Lord, and there is no other, besides me there is no God;
> I equip you, though you do not know me, that people may know, from the rising of the sun and from the west, that there is none besides me;
> I am the Lord, and there is no other.
> I form light and create darkness,
> I make well-being and create evil,
> I am the Lord, who does all these things.
>
> "Shower, O heavens, from above, and let the clouds rain down righteousness; let the earth open, that salvation and righteousness may bear fruit; let the earth cause them both to sprout; I the Lord have created it.
>
> "Woe to him who strives with him who formed him, a pot among earthen pots! Does the clay say to him who forms it, 'What are you making?' or 'Your work has no handles'? Woe to him who says to a father, 'What are you begetting?' or to a woman, 'With what are you in labor?'"
>
> Thus says the Lord, the Holy One of Israel, and the one who formed him: "Ask me of things to come; will you command me concerning my children and the work of my hands? I made the earth and created man on it; it was my hands that stretched out the heavens, and I commanded all their host."

Note that in Isaiah 45:7, God says that he creates "evil" (KJV). This word is translated "disaster" in the New International Version, and "calamity" in the English Standard Version, but the Hebrew word is the exact same word as "evil" elsewhere in the Old Testament. The translators of modern versions of the Bible (unlike the King James Version translators) avoid the word "evil" here because many theologians have argued that the "evil" which God claims to create here is "natural evil," i.e., things in nature that cause death and suffering, not human sin. Yet the creation of "natural evil" is just as problematic as the creation of evil people for young-earth creationists. They would argue that God could not create natural evil until after Adam and Eve sinned. This passage, however, flips that on its head and asserts God's right to create whatever he wants when he wants, including natural evil, independent of us.

The balance of prosperity and natural evil is strictly parallel with the creation of light and darkness in Isaiah 45:7; the rest of the passage also clearly deals with God's sovereign acts of creation, not his "response" to humans. In other words, the creation of natural evil is parallel with the very statements in Genesis 1:3–4, "And God said, 'Let there be light' . . . and he separated the light from the darkness." God asserts the sovereign right to create natural evil *in the same way* that he created light and darkness, as creation *ex nihilo*.

Natural evil, including animal death, serves as a judgment held in readiness under the covenant of law, but that is not its only purpose. Even in the absence of human sin, the destructive forces of nature show forth God's power and divine nature. As such, they are automatically judgments held in readiness, because anyone who offends God must deal with his power. Such displays of power are "very good," because the sun and the moon do not sin, nor do the land and the sea, nor do the animals. They just are what they are, and do what they were designed to do. Only when we enter in do we wish that God would not have included some of them in the universe—we stumble in the dark, we shiver in the cold, our ships sink in the sea, and we fear rats and microbes and cockroaches and sharks. If Adam had not sinned, we might have observed all of this from a protected place, in the Garden, but now we are flung among it all, still protected from the totality of God's wrath, but cursed to be "like the beasts that perish" (Ps. 49:12, 20).

We frequently see things like darkness, fire, the sea, and wild (carnivorous) animals used as judgments against sinful people in Scripture. We err, however, if we assume that these things have the exclusive purpose of judgment on sin, and therefore could not exist before Adam and Eve sinned. They existed before Adam as "potential" judgments, but they would have given glory to God even if they had never been used against humankind. Even in destroying each other, they show forth God's perfect balance and mighty power.

Should You Not Fear Me?

> Do you not fear me? declares the Lord; Do you not tremble before me? I placed the sand as the boundary for the sea, a perpetual barrier that it cannot pass; though the waves toss, they cannot prevail; though they roar, they cannot pass over it.
>
> But this people has a stubborn and rebellious heart; they have turned aside and gone away. They do not say in their hearts, "Let us fear the Lord our God, who gives the rain in its season, the autumn rain and the spring rain, and keeps for us the weeks appointed for the harvest."
>
> Jeremiah 5:22–24

But the L<small>ORD</small> is the true God; he is the living God and the everlasting King. At his wrath the earth quakes, and the nations cannot endure his indignation. Thus shall you say to them: "The gods who did not make the heavens and the earth shall perish from the earth and from under the heavens."

It is he who made the earth by his power, who established the world by his wisdom, and by his understanding stretched out the heavens. When he utters his voice, there is a tumult of waters in the heavens, and he makes the mist rise from the ends of the earth. He makes lightning for the rain, and he brings forth the wind from his storehouses.

Jeremiah 10:10–13

Some people might argue that God could demonstrate his power before the fall, but not his wrath. In general, however, it is not easy to make a distinction between the demonstration of God's power and the demonstration of his wrath in Scripture. For example, Jeremiah 5:22–24 and Jeremiah 10:10–13, quoted above, strongly connect the balance of nature in creation to the wrath *and* power of God.

These passages focus on God's creation of the world, but the terminology of wrath is seen in things like the roaring waves and the lightning. Jeremiah essentially says, "Don't you see that if God removed the balance of nature for only a moment, you would be destroyed? Therefore you should fear him who created this balance and stands as Lord of it." One of the messages we are told we should see from the creation is the threat of God's power. The logic goes as follows:

- You should fear me
- . . . because my wrath is evident to you
- . . . . . . in everything that was made in the creation of Genesis 1.

The creation of the sea should make us "tremble"; the waters in the heavens "roar" and make "tumult." These elements appear in Genesis 1:6–10, but Jeremiah tells us they should strike fear in our hearts.

This is essentially the same message that Romans 1:18–20 presents, that "since the creation of the world" God's power has been clearly seen:

> For the wrath of God is revealed from heaven against all ungodliness and unrighteousness of men, who by their unrighteousness suppress the truth.
> For what can be known about God is plain to them, because God has shown it to them.
> For his invisible attributes, namely, his eternal power and divine nature, have been clearly perceived, ever since the creation of the world, in the things that have been made.
>
> Romans 1:18–20

Looking carefully at this passage, one can ask whether the "wrath of God" that is "revealed" (1:18) is different from the "power of God" that is "clearly perceived" (1:20). Romans 1:20 clearly states that the "power of God" has been continually made known since the creation. Are we warranted to make a strong distinction between the two here? The tense of the verb in Romans 1:18 can be translated "is being revealed" (as in the NIV), which refers to a continuous, ongoing process. One can argue that the two revelations of nature in Romans 1:18 and 1:20 are the same, in other words, that God's wrath has been made known from the beginning.

To summarize, Adam was not created in an idyllic state in which he encountered only blessings. From the very start, he was confronted with two realities: blessing and curse, fruition and destruction, peace and power, light and darkness, life and death. Adam lived in a blessed, safe Garden, surrounded by an incredibly vast creation that showed forth the tremendous power of God; this power gave meaning to the word "death," which God used when he warned of the curse. If Adam had obeyed God, this "very good" creation still would have testified to Adam of the power of God and the wrath that he had escaped. As it is, Adam and all of us are flung out of the Garden to experience the same curse that was made known to Adam. All those who are outside of Christ still live by the same covenant of law as Adam, with the same visible reminders.

Why?

The biblical case seems sound enough, but we still react against the idea. Why, why? Why did God do it that way? How can God be good and make animals suffer and die? A young-earth friend of mine asks whether I can really live with the idea that animals got painful parasites and died, long before human beings ever appeared.

At one level, God reserves the right to be inscrutable. No one can fathom the reasons for all he does (Job 11:7; Ps. 145:3; Eccles. 3:11; 11:5; Isa. 40:28). We must simply trust him when he tells us it is all "very good."

At another level, however, there is at least one very obvious lesson here: God is dangerous and powerful. Paul says this explicitly in Romans 1:20: "For his invisible attributes, namely, his eternal power and divine nature, have been clearly perceived, ever since the creation of the world, in the things that have been made." We may not like the implication, but nature tells us that God is capable of severe wrath and pain. That is what his "power" includes.

This is an unpopular doctrine, but it is all through the Bible. Many times people read the Bible and are turned off by the amount of wrath in it. Not only the Old Testament, but the New Testament is filled with wrath. Some people think Jesus preached only love, but we hear more about hell from the lips of Jesus than from any other speaker in the Bible. The entire Gospel revolves around the idea of avoiding God's wrath; in fact, God pours out his wrath on his Son so that we may avoid it. The book of Revelation has page after page of wrath.

Many people say they don't believe that God is like that, but on what basis do they say that? Because nature is so gentle and kind that the God who created nature could not have done and said all those wrathful things in the Bible? On the other hand, some people who believe in the Bible say that they do not believe God would create cruel things. On what basis do they say that? Because of the complete lack of cruelty and wrath in the Bible? We have two things that agree completely—the Bible and nature—in giving us a stark picture of God's wrathful nature, but instead of accepting them, we reject both. On what basis, then, do we reject them? Merely our own wishes? If religion is about believing what is true, not just what we wish was true, then surely we must swallow the hard pill that God, the real God who exists and created the world, is not just the way we would like him to be. We may hate the wrath of God, but we cannot say it is illogical to believe in it. What is illogical is to believe in a God who would never harm a flea when we see lots of harmed fleas around us.

This may drive some people to prefer atheism, but even nature's terrors testify that God exists. We must marvel at the shark, even while fearing it. It is well designed, frighteningly so. So also are many parasites. It is hard to believe that such well-designed weapons could arise by chance—they are *good* designs. Darwinism tries, of course, to give us an explanation without referring to God, but the story of Darwin himself[2] shows that the main reason why many people *want* Darwinism to be true is that they just cannot accept the idea of God being glorified by violence. Darwin used the example of a wasp-eating larva as an example of something he just could not imagine God making in a good world.

Those who believe that all natural evils arose after the fall of Adam and Eve cannot avoid this point. If it would have been bad for God to have made wasp-eating larvae before the fall, how is it now justified? If we say that the only merit in making natural evil is to punish humans, then how are we punished by the death of a wasp? If we say, on the other hand, that the death of the wasp serves as a reminder to us of the wrath of God, why could that not have been the case *before* the fall? God's wrath did not suddenly spring into existence when Adam and Eve sinned, and God had no desire to hide this side of his nature.

The threat "in the day you eat of it, you will surely die" in Genesis 2 clearly shows that God wanted to remind people of his wrath even before the fall. Just as God's threat of wrath existed before the fall, so also the agents of wrath existed before the fall. They were not out of control, or random, but bounded carefully by God's control in an amazing balance.

If Adam and Eve had obeyed God and passed the test, they would not have stayed in the same world forever. How could they, if they were fruitful and multiplied and never died? At some point, the world with its finite surface would be overpopulated unless God took them to heaven or made a new earth and a new heaven which absorbed the old one. We don't know what God would have done in that case, but we can say that this world was not meant to last forever. There will be no wrath in heaven.

Review Questions

1. If you were a Hebrew at the time of Moses, would you have assumed that the "great sea monsters" mentioned in Genesis 1:21 were friendly herbivores?
2. What "evil" does God claim to create in Isaiah 45:7? If God creates this "evil" in the same way that he creates light and darkness, could he have created it at the same time?
3. Do Job 38–41 and Psalm 104 give primarily positive or negative portrayals of carnivorous animals? Do these passages give the picture of creatures that are "very good"?

5

The Biblical Case III

The Sabbath

I have dwelt at length on the issue of animal death before the fall because, in my experience, this is *the* issue that leads to objections to an old earth. I now turn to another issue that often drives people to a young-earth view, namely, the way in which the creation story is used in passages that deal with the Sabbath.

The Precedent of God

> Remember the Sabbath day, to keep it holy. Six days you shall labor, and do all your work, but the seventh day is a Sabbath to the Lord your God. On it you shall not do any work, you, or your son, or your daughter, your male servant, or your female servant, or your livestock, or the sojourner who is within your gates.

> For in six days the LORD made heaven and earth, the sea, and all that is in them, and rested the seventh day. Therefore the LORD blessed the Sabbath day and made it holy.
>
> Exodus 20:8–11

> It is a sign forever between me and the people of Israel that in six days the LORD made heaven and earth, and on the seventh day he rested and was refreshed.
>
> Exodus 31:17

God's Sabbath rest is used as the example for our own rest. If God did not really work six days and rest on the seventh day, how can this be an example for us? In my opinion, this is one of the better arguments against the "framework" theory, which I will discuss below, that posits that the entire chapter of Genesis 1 is figurative and that there were no creation days at all.

This argument does not immediately cause problems for the "day-age" theory, however. I will describe the day-age theory later in this chapter. The issue for day-age advocates is, "Do these passages demand twenty-four-hour days, or could they have been longer?"

It may sound trite to say that "with the Lord one day is as a thousand years" (2 Peter 3:8; see also Ps. 90:4), but we do well to remember that God's timing is not always our timing. Still, since the week in which we are to keep the Sabbath consists of twenty-four-hour days, it is certainly a natural assumption to view the days of creation as twenty-four-hour days too.

When one studies the Old Testament, however, one finds that the Sabbath law is clearly not restricted to periods of seven twenty-four-hour days. Leviticus 25 gives the Sabbath year law, as well as the Jubilee law, which was a Sabbath of Sabbath years, a period of seven times seven years:

> Speak to the people of Israel and say to them, When you come into the land that I give you, the land shall keep a Sabbath to the LORD. For six years you shall sow your field, and for six years you shall prune your vineyard and gather in its fruits, but *in the seventh year there shall be a Sabbath of solemn rest for the land, a Sabbath to the LORD.* You shall not sow your field or prune your vineyard.
>
> You shall not reap what grows of itself in your harvest, or gather the grapes of your undressed vine. It shall be a year of solemn rest for the land. The Sabbath of the land shall provide food for you, for yourself and for your male and female slaves and for your hired servant and the sojourner who lives with you, and for your cattle and for the wild animals that are in your land: all its yield shall be for food.
>
> You shall count seven weeks of years, seven times seven years, so that the time of the seven weeks of years shall give you forty-nine years. Then you shall sound the loud trumpet on the tenth day of the seventh month. On the Day of Atonement you shall sound the trumpet throughout all your land. And you shall consecrate the fiftieth year, and proclaim liberty throughout the land to all its inhabitants. It shall be a jubilee for you, when each of you shall return to his property and each of you shall return to his clan. That fiftieth year shall be a jubilee for you; in it you shall neither sow nor reap what grows of itself nor gather the grapes from the undressed vines.
>
> Leviticus 25:2–11, italics added

The fact that the word Sabbath is used here, as it is in the Sabbath-day command, shows that these commands are directly connected to the Sabbath precedent of God. Leviticus 25:4 says that the land is to have a Sabbath rest every seven years, not every seven days. Land has a different time scale for Sabbath rest than people.

If I simply argue that days of creation were not necessarily twenty-four-hour days, however, I have not really established my case. Is there a positive argument that the days of creation were not twenty-four-hour days?

Some people have pointed out that equating Genesis 2 with Day 6 implies a very busy twenty-four-hour day for Adam. Genesis 1 says that on Day 6, God created both "male and female." Genesis 2 gives a longer story of the creation of man and woman, which most conservative commentators take as an expanded version of the same story. If Day 6 was a single twenty-four-hour day, then on that one day, according to Genesis 2, Adam had to be created, name all the animals, feel lonely, fall asleep, have Eve created, wake up and find her, marry her, and receive God's commission of Genesis 1:28–30. This is possible, especially if God miraculously paraded all the animals before Adam as he called out names at high speed, but it seems to violate the sense of the text, which gives a picture of Adam carefully examining the animals and judging them for their suitability as a "helper." In Genesis 2:23, Adam says, when he meets the woman, "at last" this is bone of my bones. The "at last" in Hebrew carries the sense of "at long last," in other words, after a long time, not just a single day or half a day.

To me, however, the best argument for equating the days of creation with ages comes from the precedent of God taught by the Sabbath law. I ask, "What did God do on the eighth day?" Does God work a six-day work week, taking every Saturday or Sunday off? After his Sabbath rest, did God pick up his tools and go back to work, taking another Sabbath seven days later? If God's activity is to be taken as an exact model for us, then this conclusion seems inescapable. Yet the rest of Scripture does not support this. God's activity appears as a single continuity. In John 5:17, when questioned about why he healed on the Sabbath, Jesus said, "My Father is always at his work to this very day" (NIV), not "My Father works six out of seven days."

The Biblical Case III

In Hebrews 4:3–4, we have an apparent contradiction to Jesus's statement in John 5:17:

> For we who have believed enter that rest, as he has said, "As I swore in my wrath, 'They shall not enter my rest,'" although his works were finished from the foundation of the world.
>
> Hebrews 4:3

"His works were finished from the foundation of the world." Which is it? Is God at work or at rest, or does he work a six-day week? Hebrews 4:4 clears this up for us:

> For he has somewhere spoken of the seventh day in this way: "And God rested on the seventh day from all his works."
>
> Hebrews 4:4

Hebrews 4:3–4 clearly states that God's Sabbath rest did not end—it continues up to this day. There is no "eighth day" when God goes back to work. The work of God that Jesus refers to as continuing to this day is the same kind of work that Jesus himself performed on the Sabbath: acts of mercy and necessity, not works of generation and construction. God ceased creating on the Sabbath day and did not start again—he now works to maintain and to redeem what he created.

These passages, John 5:17 and Hebrews 4:3–4, taken together, clearly indicate a single, continuous activity of God since the creation, not a six-day work week. Hebrews 4:3–4 teaches that this entire age is the Sabbath Day of God; the surrounding verses of Hebrews 4 teach that we may "enter" that Sabbath rest, which is ongoing and waiting for us:

> So then, there remains a Sabbath rest for the people of God, for whoever has entered God's rest has also rested from his works as God did from his. Let us therefore strive to enter that rest, so that no one may fall by the same sort of disobedience.
>
> Hebrews 4:9–11

The Sabbath rest "remains," or "endures," as a state in which God himself stays, and which we can also enter into. This same rest was waiting for Adam and Eve after God had finished creating them, but they did not "enter into" it because of their sin.

If the Sabbath day is an "age," as Hebrews 4:3–4 teaches, then why cannot the other days of the creation story be "ages" also? The fact that the Sabbath day is an age seems to demand the interpretation that *all* the days were "God's days," not ours.

A good general principle is, "Scripture interprets Scripture." If all we had was Genesis 1 and 2, we might not come up with this interpretation, although we might wonder about what God did on the eighth day. The letter to the Hebrews gives us God's inspired interpretation of the Sabbath day of God.

The Day-Age vs. the "Framework" Hypothesis

As I said above, the use of the creation story as a precedent for the Sabbath does seem to strike against the "framework" hypothesis that the seven days of creation do not give any chronological information at all. The emphasis of Genesis 1 seems to be on time passing, and on its precedent for our use of time, in both Sabbath days and Sabbath years.

The framework hypothesis, associated with Bible scholar Meredith Kline,[1] says that Genesis 1 is a poetic framework for discussing the scope of God's creation. As such, the categorizing of God's creating activities under the headings of days is nothing more than an organizing principle. Because of this, we are not warranted to deduce anything about the actual sequence of events.

The day-age view is a different old-earth view. It makes the hypothesis that the days of Genesis 1 do give chronological information, but the time periods are not to be equated with twenty-four-hour days. This view has been advocated, with different variations, by old-earth authors such as Robert Newman and Herman Eckelmann,[2] and Hugh Ross.[3] In this view, the days of Genesis 1 represent long periods of time, what one may call "God's days." I will expand on this view in chapter 7; my purpose here is to compare this view to the framework view in regard to the Sabbath theme.

The best argument for the framework hypothesis is the observation that the sun and moon are created on Day 4, while light and dark are created on Day 1. How can the world have light without a sun? The framework hypothesis says that Days 1, 2, and 3 are parallel to Days 4, 5, and 6 (light and dark, sea, and dry land occur in the same order), and it argues that these are actually the same events recorded in different ways.

The question of how to get light without a sun is, of course, a problem for both the young-earth, twenty-four-hour-day view as well as for the day-age view. How can we insist on twenty-four-hour days when the clock that defines twenty-four-hour days, the sun, was not created until the fourth day? (For that matter, the moon defines the week, also, since the seven days of the week correspond to the four phases of the moon in its twenty-eight-day cycle.)

There is no doubt that the parallelism of Days 1–3 with Days 4–6, pointed out by framework proponents, exists in the text. This does not rule out the possibility that God actually created the universe in a sequence of miracles that included parallels, however. And the parallelism is not as exact as framework proponents would like. Table 4 shows the events of the different days, arranged in the parallel form favored by framework proponents.

Day 1 light, darkness	Day 4 sun, moon, and stars
Day 2 heaven (sky)	Day 5 sea creatures, flying creatures
Day 3 land and sea, plants	Day 6 land creatures, humans

Table 4. The days of Genesis 1 arranged in parallel.

The framework interpretation argues that Days 4–6 present the things that populate, or rule, the first three days, and therefore may be taken as describing the same events. As one can see from this table, however, sea creatures are created on Day 5, but to be perfectly parallel, should have been created on Day 6. It also would seem to be more parallel if the stars were created on Day 2, as the population of the nighttime sky.

A framework proponent might argue that the sea is created on Day 2, as the waters below the heavens, but the days are defined by the things that God calls by name. In Day 1, God calls the darkness and the light by name; in Day 2 he calls heaven by name, and on Day 3 he calls the land and the sea by name. The waters under the heavens in Day 2 are a mixed state of land and sea, waiting to be separated.

The day-age view takes the chronology of Genesis 1 seriously: the days of Genesis 1 represent a real sequence of seven ages. In this view, the Sabbath precedent is also taken seriously. The Sabbath precedent is problematic in the framework view, however. The Bible says over and over, "God rested on the seventh day."

The Sabbath in the Book of Revelation

If the days of Genesis 1 are not literal twenty-four-hour days, one may ask why God would talk about a single, long period of time as a sequence of seven ages. The answer is that the number seven always represents perfection and completeness in the Bible. The word for Sabbath in Hebrew is a very broad concept that carries with it the concepts of wholeness, rest, and blessing. As argued above, Hebrews 4:3 indicates that we are in a "seventh age" of God, the age in which he is at rest from all his creative activity, though not his redemptive work. This age follows six ages of creation, defined by God, for a perfect whole.

The use of the number seven to indicate completion or perfection also appears strongly in the book of Revelation. The symbolism of the seven days of creation has a close analogy with the seven seals introduced in Revelation 5–6, along with the seven trumpets of Revelation 8–9 and the seven plagues of Revelation 16. These sequences of seven describe a progressive unfolding of the destruction of the world, a sort of "anti-creation week." God could simply end everything at once, if he wished. Yet he chooses, as in the creation, to not do everything at once. Revelation shows a process of de-creation over time.

There are many different views of Revelation, of course, but even without committing to a specific school of thought regarding the events described, one can notice the structure. In each cycle of seven, after six stages of destruction, the seventh stage brings rest:

> When the Lamb opened the seventh seal, there was silence in heaven for about half an hour.
>
> Revelation 8:1

> Then the seventh angel blew his trumpet, and there were loud voices in heaven, saying, "The kingdom of the world has become the kingdom of our Lord and of his Christ, and he shall reign forever and ever."
>
> And the twenty-four elders who sit on their thrones before God fell on their faces and worshiped God, saying, "We give thanks to you, Lord God Almighty, who is and who was, for you have taken your great power and begun to reign. The nations raged, but your wrath came, and the time for the dead to be judged, and for rewarding your servants, the prophets and saints, and those who fear your name, both small and great, and for destroying the destroyers of the earth."
>
> Revelation 11:15–18

> The seventh angel poured out his bowl into the air, and a loud voice came out of the temple, from the throne, saying, "It is done!"
>
> Revelation 16:1

Both Genesis and Revelation use seven periods to indicate a *long process*, and the number seven indicates *completion*. Some commentators see the events of Revelation as taking all of history from the time of the apostles until now, while others see them as spanning just a few years, but I don't think that any commentators insist that the sevens in Revelation represent seven twenty-four-hour days. Given the close analogy of the creation week and its Sabbath with the de-creation sequences in Revelation, it is therefore reasonable to expect that if the events of Revelation take longer than 168 hours, then so might the events of Genesis 1–2.

I am not saying that Genesis 1 is written in the same apocalyptic style as Revelation. Rather, I am saying there is a deep significance to the idea of seven sequential events in the Bible, and that these events may be used to describe a period of indefinite length, both seven calendar years, as in the case of the Levitical Sabbath-year law, or indeterminate periods, as in the book of Revelation.

An advocate of the framework hypothesis might argue that both the days of creation and the cycles of seven in Revelation are symbolic, not really recording time sequences. The first five seals, for example, represent things that could happen at any time in history: tyranny, war, famine, death, and martyrs. The sixth seal, however, seems to indicate events that happen after the other events, an acceleration of judgment:

> When he opened the sixth seal, I looked, and behold, there was a great earthquake, and the sun became black as sackcloth, the full moon became like blood, and the stars of the sky fell to the earth as the fig tree sheds its winter fruit when shaken by a gale. The sky vanished like a scroll that is being rolled up, and every mountain and island was removed from its place.
>
> Revelation 6:12–14

The later cycles of seven give an even stronger indication that a sequence of events is in view. For example, in the cycle of the seven trumpets, the fifth and sixth trumpets begin periods in which special events occur: demonic locusts coming from the pit of hell (Rev. 9:10–11) and the release of millions of horse-serpents (9:19), respectively. The angels' words regarding these events indicate sequence, as well. In Revelation 8:13, just before the fifth trumpet, we read

> Then I looked, and I heard an eagle crying with a loud voice as it flew directly overhead, "Woe, woe, woe to those who dwell on the earth, at the blasts of the other trumpets that the three angels are about to blow!"

Then in Revelation 9:12, before the sixth trumpet, we read,

> The first woe has passed; behold, two woes are still to come.

And in Revelation 11:14, before the seventh trumpet,

> The second woe has passed; behold, the third woe is soon to come.

Additionally, in the cycle of the seven bowls of plagues, events of one bowl are undone in the next—the fourth bowl brings a scorching sun, while the fifth bowl brings darkness (Rev. 16:8–10). All of these time-cues indicate that real chronology is being presented.

The cycles of seven also appear sequential relative to each other. The seven bowls come as the sevenfold completion of the seventh trumpet; the seven trumpets come as the sevenfold completion of the seventh seal. The seven seals, one may argue, themselves come as the sevenfold completion of the Sabbath day of creation. Thus the events of the seven seals represent the "beginning of the birth pangs" mentioned by Jesus in Matthew 24:4–8. With this view, events such as wars and famines and martyrdom are not the end yet, but only the beginning of the end. The events of the seven trumpets, with their repeated emphasis on "one third" of things being destroyed (Rev. 8:7–12; 9:15), represent partial destruction of the world. They take place after the "day of their wrath" has already been proclaimed (Rev. 6:17), and the events of the seven bowls represent total destruction of the order of the world—"Then I saw another sign in heaven, great and amazing, seven angels with seven plagues, which are the last, for with them the wrath of God is finished" (Rev. 15:1).

The objection of framework proponents to the day-age hypothesis sometimes sounds like Augustine's objection to seven days of creation. In the fourth century, Augustine was not concerned with why the time of creation should be so short as seven days, but why it should be so long. Why should God, who could create everything instantaneously, take as long as seven days, even twenty-four-hour days? Augustine proposed a variation of the framework view in which the days of creation are essentially symbolic. The modern variation is to say that seven definite time periods is somehow below God; he should do it in one seamless process without breaks. Yet who are we to tell God whether he should do it instantaneously, in one continuous process, or in seven stages?

Similarly, some people today think it strange to describe the final judgment as a long process; it seems much simpler to simply say that Jesus comes back and all are judged in one event. Yet the book of Revelation gives us a more detailed picture, with a number of events in sequence. The events of Revelation are arranged in weeks with Sabbath rest, in periods of indeterminate length. If we take seriously the sequences of Revelation as representing a real chronology of events over a long period of time, then it is natural to see a parallel with the sequence of Genesis 1 representing a real chronology over a long period of time.

Marking the Days

One version of the day-age view maintains that the beginnings and endings of the ages would be impossible for us to demarcate exactly by external, physical signs; they are exclusively "God's days," new stages in the development of the world without sharply defined boundaries. Another possible view, however, is that each age begins with a special, divine miracle. As discussed in chapter 2, an old-earth view is not synonymous with theistic evolution. Scientists in the intelligent design movement would say that even billions of years is not long enough to make evolution probable without either divine "rigging" of the laws of nature or miraculous divine intervention. One view of Genesis 1, then, takes specific events of divine intervention as markers of the ages of creation. I will return to this idea in chapter 6. At this point I simply ask, why not suppose that seven miraculous events marked the beginnings of the ages? The Bible is a story of miracles. Whether we can identify these miracles using modern science is a secondary issue.

The day-age model declares that the sequence of days gives real chronological information, a precedent for our own use of time, but it takes God's days as typological of our own, not strictly equal to them. This view allows us to take the Sabbath precedent used in the Sabbath laws seriously, and provides for a natural understanding of the Sabbath of God discussed in Hebrews 4. It also highlights a parallel with the sequences of seven in the de-creation of Revelation. If there is no Sabbath age following the six ages of creation, then we seem to be forced to the odd conclusion that God works only six days out of every seven days.

Review Questions

1. Does Hebrews 4:3–4 teach that God's seventh day is ongoing? If so, then if the Sabbath day is an "age," could the other days be ages also? If not, then does God have a Sabbath every seventh day? Why is there no mention of an "eighth day"?
2. If God's precedent in the Sabbath must be the same length of time as ours, then what is the precedent

for the Sabbath year and the Sabbath of Sabbaths, or Jubilee? If God's Sabbath is typological of those, then why not the work week also?
3. Can we insist on twenty-four-hour days when the clock that defines twenty-four-hour days, the sun, was not created until the fourth day?
4. Is the parallelism of Days 1–3 and Days 4–6 exact enough to support the framework hypothesis? Why are sea creatures created on Day 5 when the sea is created on Day 3?
5. Do the seven periods of "de-creation" in the book of Revelation consciously parallel the seven days of Genesis 1? If so, are the periods of judgment in Revelation twenty-four-hour days?

Concordantist Science 6

In chapter 1, I argued that science may sometimes affect our interpretation of the Bible. There, I mainly argued against the "fundamentalist" who would seal off the Bible from the rest of our experience.

In this chapter, I argue that our understanding of the Bible can also affect what we look for in the world of science. Here I argue against the "liberal" who would seal off the world of our experience from the Bible. I am arguing the same way I did in the first chapter: the Bible and science do not exist in two, non-overlapping worlds. If we allow them to speak to each other, the flow can go both ways; therefore, it is natural to use information from the Bible to direct our inquiries in the scientific arena.

In the previous chapter, I discussed the day-age view, which says that we can line up creation days with our modern scientific understanding of geology and archaeology. This is known as "concordantist" science. In chapter 7, I will give my own version of how this works out; Newman and Eckelmann,[1] Hugh Ross,[2] John Wiester,[3] and G. Schroeder[4] present other versions.

At first, this seems impossible, since Genesis 1 has, for example, the plants appearing before the sun and the moon, while the modern understanding is that the sun and moon appeared long before plants. Proponents of the day-age view, however, have argued for an "observer-based" understanding of the appearance of the sun and moon on Day 4. Just as we allow that phrases like "the sun rose" may be taken as true from the perspective of an observer on earth, so day-age proponents argue that the command "Let there be lights in the sky" refers to the fact that sun and moon did not appear *in the sky* before then, as viewed from the surface of the earth. The "sky" as used in Scripture does not refer to outer space, but to the atmosphere above us, as seen by a person on the ground. Since no one had traveled in space then, what else could it refer to meaningfully? Day-age proponents argue from modern science that the clouds parted on the earth at a definite time in history, utterly changing the ecology of the world, and that this change occurred after the green plants had appeared and before animal life. The fact that our atmosphere is transparent is actually quite surprising;[5] the most likely state of the earth is to have a cloudy atmosphere like the other planets.

A number of modern authors mock concordantist science as a misguided project. Some argue that it is unwarranted to see any scientific information in Genesis because it is a religious book, not a scientific one. At first blush it does sound silly to think of God slipping information about interstellar plasmas into a book written to ancient Hebrews who lived in tents and traveled on camels.

But while it is true that Genesis is not a scientific textbook, one can not so easily separate the world of science from the world of religion without making an unnatural division in the world that the Bible itself does not make. Although the main purpose of the Bible is not to teach science, science is involved in everything we do, since science is nothing but a way to organize and analyze the things of the world around us. And the Bible goes to great lengths to emphasize that it deals with the real world around us.

Francis Schaeffer stressed this point constantly in his books, coining the term "true truth" to describe it—the Bible talks about our world, the one where we live, not some mythical land of allegory. For example, in Genesis 2:8–14, the Garden of Eden is placed in the real world with the names of real rivers and nations. People have genealogies, lands are given place names, dates are tied to the reigns of real kings. We may be bored by such details, but they serve the important purpose of reminding us that these people really existed. If we forget this, we end up making the Bible merely an interesting set of moral allegories instead of what the Bible really is: a historical record of the intervention of God into our real world.

Because of this, innumerable things in the Bible are open to scientific investigation. We are all familiar with archaeological digs that seek to confirm facts recorded in the Bible; biologists have also looked into the flora and fauna described in the Bible. Looking to line up geology and cosmology with the Bible is no different.

We can make an analogy here with the Levitical dietary laws. Many scientists and medical doctors have noted that obedience to the dietary laws of Moses leads to good health. By modern health standards, many of these laws seem obvious: don't eat road kill, don't eat animals that eat their own dung and vomit, don't eat creatures that suck scum from the bottom of standing water, don't touch dead people unless you wash afterwards, dispose of excrement outside the city. These laws concord very well with what we now know about bacteria and the spread of disease. The intent of these laws was not to convey information about bacteria. Instead, they served as a pattern of cleanness that symbolized the separation of God's people from the rest of the world. Yet underlying them is valid scientific truth, which the people could never have known at the time. God, being sovereign, could make his laws consistent with things that only modern science could discover; while the people then did not need to know the science behind the laws, our knowledge of it gives us a richer understanding of those passages today.

If we interpret Genesis 1 in terms of modern geology and planetary science, this does not mean that we assume that God was hiding detailed science in a text to ancient Israelites who would have had no clue about such concepts. Such a view would make us akin to the Gnostics, trying to find secret meanings instead of taking the text at face value. God did not hide anything. He simply conveyed basic ideas in a picture language that was nevertheless true and accurate. Just as he said, on occasion, "the sun rose" (e.g., Gen. 32:31; Jonah 4:8), knowing full well that the earth rotates, so he could say "the sun appeared in the sky" without discussing planetary formation.

But What If We Are Wrong?

What if we find a particular way of lining up geology with the days of Genesis 1, and then the scientific world changes its view of how the earth formed? Many Christians seem to be afraid to make any predictions based on the Bible that could be falsified, for fear that people will reject all of Christianity if it is attached to a particular scientific theory. This, in my experience, lies at the root of much of the objection to concordantist science. Many Christians want to seal off their Christian belief from any possible contradiction with science, so that it is an impregnable fortress against all attack.

I call this basic mind-set of so many modern Christians, both conservative and liberal, the "two-worlds" view.[6] This view says, in essence, that science and real-world experience lie in one world and that the Bible and theology lie in another world, completely distinct from the first. The two worlds do not contradict each other because they cannot; no overlap exists so one world does not have implications in the other. The Bible has authority in matters of faith, but not at all in matters of science, because faith and science have nothing to say about each other. People use various terms such as "orthogonal," "complementary," and "different levels of description" to describe this non-intersection of worlds.

This two-worlds mind-set reflects an essentially defensive posture. Having survived a long tradition of attack on Christianity in the name of science, many Christians assume that if the two worlds did overlap, then science would surely contradict Christian faith. Even if science does not presently appear to contradict our faith, the possibility always exists that it will.

In the early part of the twentieth century, Christianity seemed constantly on retreat before the victories of the modernist movement. Darwin showed the Bible was wrong on origins, Freud showed the Bible was wrong on guilt, archaeology showed the Bible was wrong on places and times, and even the miracles of Jesus were explained away. In response, Christians turned to existentialism and postmodernism, which essentially say that we can "leap" to belief in any value system we want. The Bible, in this approach, is no longer a report of events in the world of science and experience, but a book of morality and "values." In so doing, they protected Christianity, but they also made it irrelevant. People may choose to leap to other belief systems, and there is nothing we can say to convince them otherwise. Once we stop thinking that the Bible deals with facts about the real world, we have no common ground on which to say that our God has intervened in their history.

More recent science has shown that the triumphalism of the modernists in the early twentieth century was premature. The intelligent design movement has undermined Darwin; Freud is largely discredited; archaeology constantly reports new confirmations of biblical stories. Suppose we did not live in such times, however, and all science seemed to contradict the Bible. Do Christians have the option of adopting an unchanging theology that knows only the Bible and not the latest scientific data? To put it another way, can we ever put our faith in such a safe place so that no datum of experience could ever overturn it?

In 1991, I wrote an article[7] presenting an imaginary scenario in which scientists claimed to have discovered the bones of Jesus, thus contradicting the claim of the resurrection. This idea was later incorporated into the plot of a popular novel.[8] I envisioned two responses to this claim. One group of Christians said, "I reject that report because it comes from scientists." This is the "fundamentalist," or anti-intellectual, position—the scientist as villain. The second group of Christians said, "I can handle that. The essence of the Bible's teaching about the resurrection does not center on the fact that Jesus really rose bodily." This is the "eager-to-please" position—the scientist as god. Both of these positions maintain a sharp division between the truth of the Bible and the truth of science. In the first case, the truth of science is unimportant, while in the second case, the literal sense of the Bible is unimportant.

No one these days worries about scientists discovering the bones of Jesus, although scientists recently claimed to have discovered the burial container of the bones of James, the brother of Jesus. But the centrality of the resurrection in Christianity tells us that our faith is inherently risky: it depends on certain historical facts that we trust are true. As Paul said,

> For I delivered to you as of first importance what I also received: that Christ died for our sins in accordance with the Scriptures, that he was buried, that he was raised on the third day in accordance with the Scriptures, and that he appeared to Cephas, then to the twelve. Then he appeared to more than five hundred brothers at one time, most of whom are still alive, though some have fallen asleep. Then he appeared to James, then to all the apostles. Last of all, as to one untimely born, he appeared also to me.
>
> 1 Corinthians 15:3–8

> But if there is no resurrection of the dead, then not even Christ has been raised. And if Christ has not been raised, then our preaching is in vain and your faith is in vain. We are even found to be misrepresenting God, because we testified about God that he raised Christ, whom he did not raise if it is true that the dead are not raised. For if the dead are not raised, not even Christ has been raised. And if Christ has not been raised, your faith is futile and you are still in your sins.
>
> 1 Corinthians 15:13–17

Paul does not argue merely for "values" or ethics based on allegorical stories; he argues on the basis of testimony about events in the real world. He leaves us open to risk: if Jesus did not rise from the dead, our faith is in vain. If the Pharisees had produced the bones of Jesus in the first century, Christianity would not exist.

An analogous split between camps occurs regarding the creation story. One group dismisses any input from science, adopting a young-earth creationist view even if all science says otherwise, and assuming that most scientists are either villains or brainwashed idiots. Another group says that we must interpret Genesis entirely figuratively, adopting whatever science tells us without concern for any implications the Bible texts may have for the real world.

In my story of Jesus's bones, a third believer, the "Seeker after Truth," says, "That really goes against the Bible, and I don't believe it. But you seem to have built a compelling case, so I want to examine this further. I expect that your science has errors, in which case I can advance science by discovering them. If your claim truly is airtight, however, then my faith has no basis, and I can not take that possibility lightly." This is a much more difficult position than the other two, each of which easily resolves any tension between science and the Bible. The concordantist with the day-age view of Genesis takes this more difficult middle road. The Bible seems to give chronological information. Rather than dismissing it out of hand, or dismissing science out of hand, the concordantist questions assumptions made both in Bible interpretation and in science in an attempt to see harmony between the two.

We must face the facts: if the Bible is wrong, utterly wrong, about the history of our origins, then we should dump it. We cannot avoid this risky aspect of our faith. If we protect the Bible by attacking modern science, or if we protect it by making it speak only about matters of morality and personal faith, we have cut it off from the real world and made it far less than it claims to be.

God of the Gaps

As discussed above, many Christians seem to be afraid to make any scientific predictions based on the Bible that could be falsified, for fear that people will reject Christianity if it is attached to a particular scientific theory. But why should that be so? If an atheist makes an incorrect prediction, or if someone says that atheism is supported by a scientific theory that later turns out to be false, no one says that atheism has been falsified once and for all. Why should Christians be afraid to use biblical information in their science?

A standard way of dismissing both intelligent design theory and concordantism is to say that they are "God-of-the-gaps" approaches. God-of-the-gaps has become, in our day, an automatic put-down of any attempt to say that science actually supports Christian belief.

The standard story of the God-of-the-gaps argument is that for centuries, Christians argued for the existence of God on the basis of things we did not know in the area of science. As science progressed, these "gaps" narrowed, leaving fewer and fewer places to put God. Any attempt to see God in science is therefore doomed by the inevitable progress of science.

This story is, to put it bluntly, a myth. There never was any serious Christian attempt to prove the existence of God based on lack of knowledge in the area of science. There have been three main philosophical lines of argument for the existence of God. The first, associated with Aquinas, argues on the basis of philosophical concepts about the whole universe, for example the argument from cause and effect. The second, associated with Calvin, argues on the basis of our inner sense of God's presence and voice. The third, associated with the British theologian William Paley, argues on the basis of the beautiful design we see in nature. All three of these arguments are alive and well today, with the same strengths and weaknesses that they have always had.

We have all heard stories of people who thought that comets were signs from God, or that magnets were works of spirits, etc. But did these people argue for the existence of God based on such things? Of course not. They assumed the existence of God, and explained some phenomena based on what seemed the most natural explanation at the time, given the existence of God. In the same way, atheists have come up with theories that seem silly now, such as Lamarkian evolution and spontaneous generation. When science moved on and discredited these theories, no one said that atheism was being pushed into smaller gaps. Atheism does not demand that such theories be true, although it is helped by any credible story of how life could have come about by mindless forces. Likewise, theism was not shaken when scientists learned about the orbits and composition of comets, though some people in earlier times might have felt comforted by the idea of regular miraculous communications from God.

There are two ways to involve God in science. One is to work within the Christian worldview, assuming God exists, and ask how God might have worked. Early medical workers in Europe tended to work this way, assuming that God had designed the body; those who saw comets as signs from God also worked this way. I will call this a "Type 1" interaction. Another way to involve God in science, which I will call a "Type 2" interaction, is to engage in evidential apologetics, working within a *non*-Christian worldview to ask about the things that atheism does not have a reasonable explanation for. In effect, one makes an argument of *reductio ad absurdum*: starting with atheistic assumptions, one finds facts from the world of science that are accepted by the person making those assumptions but which seem to contradict the reasonable expectations of someone holding an atheistic view. In elementary logic, if deduction from a set of assumptions leads to a contradiction, at least one of the assumptions must be false.

This "Type 2" type of argument is known as a "Bayesian" argument. It is not an all-or-nothing proof. An atheist might agree that there is no good explanation for some things within his or her worldview, and just hold out for a future explanation. In the same way, a Christian might agree that there are some things without good explanation in the theistic worldview, and not feel threatened, holding out for more information. Just because an evidential scientific argument does not automatically cause the capitulation of the other side does not mean that such arguments are useless. Evidence that seems to contradict a worldview can "weaken" that worldview; if enough evidence mounts that is inexplicable according to a preconceived view, a person may reach the breaking point and switch to a new view, something that is often called a "Kuhnian revolution."

Intelligent design theory is this kind of "Type 2" science-faith interaction. Concordantist day-age theory is a "Type 1" science-faith interaction, an attempt to make sense of the world within the Christian worldview. A concordantist day-age theory that successfully integrates science with Genesis 1 will not convince any atheists to become theists, and its failure will not cause Christians to give up on the Bible. A successful concordantist day-age theory, however, like a successful model of evolution for the atheist, would provide Christians with a coherent picture of origins, and remove the confusion that arises from things in science which seem to contradict basic biblical premises.

The Question of Miracles

As mentioned at the end of chapter 5, there are two versions of day-age theory. One approach says that the days represent ages which are God's days, with no particular markers in our world. The other approach posits that each day is marked by a different type of miraculous work by God.

In using the word *miracle*, I will offend some people. Some Christian authors, such as Howard Van Till,[9] dislike the idea of introducing miracles into the development of the universe because this seems too undignified for God. This is akin to Augustine's objection to God using seven whole days to create the world—it seems too undignified. Van Till feels that if we say the development of the universe was not seamless and continuous, we accuse God of incompetence, of needing to fix things up *ad hoc*. But Van Till himself believes in the miracles of the New Testament, such as the feeding of the five thousand and the resurrection. Why should one type of miracle be undignified and messy, and not the other?

Many people think that believing in miracles is incompatible with science, that science must always seek to find a non-miraculous explanation for things. This approach is sometimes called "methodological naturalism." The methodological naturalist argues that if we allow a miraculous explanation, we will give up too soon looking for natural explanations. Authors like Van Till think that those who believe in miracles will just say "it's a miracle!" and stop doing science whenever they face something they do not understand.

Will it blunt our curiosity or make us give up too soon if we believe something has come about by a miracle? This does not follow. As I mentioned above, the roots of modern medicine and biology sprang from Christians who believed that the body was well designed. This did not stop their curiosity; rather, it drove their curiosity—they wanted to find out how God had done things. In the same way, early scientists studying cosmology felt they were thinking God's thoughts after him.

By analogy, imagine an engineer given two computer chips. For one chip, the engineer is told, "this chip was designed by a master designer." Regarding the other, the engineer is told, "This was assembled by a random process and has large parts that have no function." Which will the engineer want to spend more time studying? In fact, engineers in industry do spend a lot of time "reverse engineering" other people's designs. Believing that something was created by an intelligence does not destroy curiosity, it increases it.

Many philosophers have tried to define science in one way or another. They imply that things that do not conform to that definition do not count as science, and therefore can be ignored. As a practicing scientist, I am not interested in whether what I do conforms to someone's definition of science. I am interested in one simple thing: finding out what is true. If it is *true* that a miracle happened, then saying, "But that isn't science" is irrelevant.

This doesn't imply gullibility. As with any truth claim, I want to see evidence. If someone claimed a miracle had happened, I would start by looking at whether the event could be explained more easily by natural processes or human deceivers. But since I believe in God, I allow for the possibility of a miracle. Something which would require a fantastic conjunction of improbable coincidences to occur naturally would be easier to interpret as a miracle. So, for example, if I had been around when Jesus healed a man born blind, I might have wanted to interview his parents, and to look at his eyes. At some point, though, it would have been irrational to continue to hold out against a miraculous explanation at all costs, imagining that all of the people involved were liars and conspirators, or that natural processes had healed him just at the right moment.

It is not good science to force an interpretation of the facts, whether natural or miraculous, at all costs and despite all evidence. A Christian working from the day-age perspective has the luxury of being open-minded about events. God may have used natural means in many cases to create the world as we know it. But in many cases he may not have. The text of Genesis 1 leads us to expect, though not force, evidence of miracles in history. Just as science can lead us to question our interpretation of the Bible, the Bible can lead us to look for things in science we might not have looked for otherwise.

Candidate Miracles

In a sense, every miracle can be described as a "singularity." Science can describe everything up to and after the event, but the event itself is beyond the bounds of science; in fact, science itself implies that science breaks down at that point. As discussed above, some Christian writers think singularities are beneath God's dignity, that it is more honorable for him to do things by continuous processes. If singular breakdowns in the laws of nature happened randomly and frequently, that might be the case. But the miracles of the Bible always come on command, in response to the word of God or his prophets. Like a king with a well-run government, God is glorified both by the smooth operation of the laws in daily life, and also by a swift response to his new and different commands.

With this in mind, let us look at some possible singularities seen in science. The first and most obvious candidate is the Big Bang.

Many concordantist authors take the view that God's statement "Let there be light" corresponds to the Big Bang. I do not take this view, as I will discuss in chapter 8; instead, I set the Big Bang at the very first line of Genesis, "In the beginning God created the heavens and the earth"—namely, all that there is.

It is unfortunate that some people oppose the Big Bang theory out of the feeling that it implies randomness in creation, because nothing could be further from the truth. Perhaps the name leads people to the wrong picture, of a chaotic explosion. On the contrary, modern science[10] has shown that incredible balances and tuning were involved in the Big Bang, with precisions on the order of one part in 10^{100} (ten followed by one hundred zeroes). Without that fine tuning, planets and stars could not exist, and the universe might not even have existed for more than a few milliseconds.

It pains me that numerous Christian science textbooks mock the Big Bang as an atheist conspiracy. As Hugh Ross has emphasized in *The Fingerprint of God*, the Big Bang theory was not a product of atheist propaganda. It was, in fact, rejected by nearly all atheists, accepted only when the burden of evidence in favor of it became overwhelming.[11] Atheist philosophy demands an eternal universe, since, as R. C. Sproul has eloquently reminded us, logic demands that something cannot come out of nothing.[12] A Big Bang is a beginning. Modern atheistic cosmology has attempted numerous theories to explain away this beginning, using adaptations of quantum theory or inflation theory to argue that ours could be one of many universes to pop into existence,[13] but these theories of multiple universes, unlike the Big Bang theory, have no observational support and are merely wishful thinking.

Many people have a basic misconception about the Big Bang theory. It does not say that all matter was concentrated at one point, surrounded by empty space, and then it blew up, expanding into the empty space. Instead, the Big Bang theory says that *space itself* was concentrated at one point, and outside it was nothing—no space at all—and then space itself rapidly expanded. In other words, the Big Bang theory is a theory of something out of nothing—appearance of space *ex nihilo*. This is hard for us to visualize because we cannot visualize "nothing," and it is hard to imagine empty space getting bigger.

Where observation is lacking, theories proliferate, limited only by our imagination. Scientists can imagine all kinds of theories of how the Big Bang could actually be a tunnel to another, pre-existing universe, or a bubble in a larger universe—if, that is, we absolutely must have a theory without God. In the absence of any compelling observational evidence for such theories, though, the Christian should have no problem with identifying this event as the creation point of the universe: a miracle.

Another candidate miracle is star and planet formation. We all have seen convincing pictures of planets forming from clouds of interstellar gas in space, but their existence is amazing: the density of our planet is 28 orders of magnitude (1 followed by 28 zeros) larger than the average density of the universe. This requires compression and cooling far beyond simple contraction of a gas cloud. The original formation of the stars and planets is still difficult to explain in terms of modern science. There is a very simple reason why: although the force of gravity increases as $1/r^2$, where r is the radius of a cloud of gas, the outward centrifugal force increases as $1/r^3$.[14] Therefore, unless a star can get rid of angular momentum through some exotic means such as magnetic fields, star formation is impossible. Although astronomical observations indicate that star formation occurs in the universe now, it is entirely second-generation star formation, in which existing stars play a role in the birth of new stars. There is presently no comprehensive model for how the first stars and planets came about.

Another event considered miraculous in many concordantist interpretations is the Cambrian explosion, a relatively short time (on geological time scales) in which biological diversity in the fossil record suddenly appears.[15]

One criticism of this interpretation is that the Cambrian explosion, while short on geological time scales, could easily have lasted several million years, which does not seem to fit the idea we have of a sudden miracle. Some have also argued that great diversity of life could have existed before then, but if the creatures did not have hard shells or bones, we would have no record of them. Nevertheless, the incredible diversity of forms that appears at the Cambrian is amazing.

Last, modern humans seem to "explode" on the archaeological scene around thirty thousand years ago. Many concordantists, naturally, equate this time with the creation of humans. There is great debate, however, over what makes a modern human. Archaeological evidence indicates that human-like creatures as far back as a million years ago buried their dead, made tools, and collected pretty things. Of course, we see animals today doing these same things. Elephants mourn their dead, monkeys use tools, and birds collect colorful objects. Despite the similarities, the evidence still indicates that "modern" humans appear suddenly, with greatly expanded brain capacity, language, and culture.

My point in all this is not to make a case for a specific identification of these events with the events of Genesis 1. I am simply arguing that it is neither unscientific nor unbiblical to try to line them up that way. The Bible and science do not lie in separate, non-overlapping worlds; therefore it is natural to look for concordance between the two.

Coming up with a reasonable day-age model involves both scientific study and Bible interpretation. The appeal of this view is that it takes the chronological phrases of Genesis 1, the passages in the rest of the Bible that talk of the creation week, and the Sabbath precedent seriously. We may be able to line up the days of Genesis 1 with specific events that science tells us happened in the past, or we may have to leave the exact demarcation of God's days as a mystery known only to God. There is no reason not to look for such correspondence, however.

Review Questions

1. What is the significance of the extensive genealogies and descriptions of geography in the Bible, in particular in the book of Genesis? If the book of Genesis were removed from its temporal and spatial setting, would its meaning change?
2. Can we ever put our faith in a place where no observational evidence could ever threaten it? Was that a goal for the apostles when they preached the resurrection of Jesus?
3. Is it legitimate to use modern medicine and science to see new meaning in the laws of Leviticus?

Interpreting Genesis 1 and 2

Is it ever possible to read a passage of Scripture with complete objectivity, removing completely the influence of our own culture? Most evangelical Christians agree that the proper understanding of any passage of Scripture is the one in which we take the words to mean what they would have meant to a reader in the day they were written. If we take other meanings, we can end up like the medieval mystics who read all kinds of fantastic meanings into the words of Scripture. But recovering the original meaning is not so simple. We cannot strip ourselves of all our cultural background and become blank slates.

This is a thorny problem no matter which side of the debate about the age of the earth you come down on. While old-earthers are often accused of bringing in scientific knowledge that was not available to the Hebrews, young-earthers and flood geologists need to be aware that they may also bring in modern notions, such as the idea of a global earth. Let us try as well as we can to imagine ourselves in the place of an ancient Hebrew reading Genesis 1, and see if we can recover a reasonable interpretation.

The Curtain Rises

Before we begin, I have one note about translation: it is simply incorrect to demand that the word *eretz* means "earth" in the sense of a spherical globe. The Hebrew word simply means "land," with all the vagueness inherent in the English word. What would land have been in the mind of an ancient Hebrew? The land he was in, from horizon to horizon. Let us imagine ourselves standing in *that* land, as the story opens.[1] In the following passages, I use the word *land* in place of *earth*, as is perfectly proper given the original meaning of the word.

> In the beginning, God created the heavens and the land.
>
> Genesis 1:1

Almost all commentators agree that this is a sweeping, all-inclusive statement, essentially saying, "God created everything you can see." The land and the heavens—all that the eye can see. The fact that the word "heavens" is plural has been taken by some people to mean that a spiritual realm is in view. This is probably pushing the text too far; the word "waters" that appears in the next verse is also plural, but no one takes this as referring to a spiritual realm. In general, the plural in Hebrew is often used for things that are great and mighty without indicating a separation into different parts. It is enough to say that the term "the heavens and the land" refers to "everything as far as the eye can see." The spirit world is simply not in view here, although that does not imply that the spirit world was not created by God.

> The land was without form and void, and darkness was over the face of the deep. And the Spirit of God was hovering over the face of the waters.
>
> Genesis 1:2

Many pages of ink have been spilled over the meaning of this verse. One view, which I do not agree with, is the "gap" hypothesis,[2] which says that between Genesis 1:1 and 1:2, the earth has been created, a major battle of angels and demons has occurred, and the earth has been left desolate, making the rest of the story of Genesis 1 a story of "re-creation," or repair of the damage done by the demonic wars. Like the hypothesis that the curse led to a re-creation of the whole world, this view inserts an entire history in the middle of a seamless narrative. It also inserts a discussion of angelic history into what appears to be entirely about the physical world. We occasionally get glimpses into the doings of the spirit world in the Bible, but those are few and brief. The beginning of the Bible is concerned about the physical land of the Hebrews, and it stays focused on that physical history up through the stories of Abraham, Isaac, and Jacob.

The people who promote the gap theory use as evidence the fact that the words "without form and void" occur also in Jeremiah:

> I looked on the land, and behold, it was without form and void; and to the heavens, and they had no light. I looked on the mountains, and behold, they were quaking, and all the hills moved to and fro. I looked, and behold, there was no man, and all the birds of the air had fled. I looked, and behold, the fruitful land was a desert, and

all its cities were laid in ruins before the LORD, before his fierce anger. For thus says the LORD, "The whole land shall be a desolation; yet I will not make a full end. For this the land shall mourn, and the heavens above be dark; for I have spoken; I have purposed; I have not relented, nor will I turn back."

Jeremiah 4:23–28

There is no question that there is a parallel with Genesis 1 here, but the picture is of destruction so severe that it will essentially de-create the land. God is predicting a severe judgment (which some may believe is the final judgment, and others may associate with a war during the days of Jeremiah) that will so devastate the land that it will return to its Genesis 1:2 state. There is no implication that Genesis 1:2 must represent the aftermath of a war.

Notice, though, that in Genesis 1:1–2 we already have all the elements of the world of the ancient Hebrew: the heavens, the land, and the waters. The concordantist day-age view of Ross and Newman and others would associate these with the plasmas of the early universe, but this would not only be outside the ken of the early Hebrews, it also seems to stretch the standard concrete meanings of the words "heavens," "land," and "waters."

I conclude that in Genesis 1:2, the point of view is that of a person on the earth, in the land of the Hebrews. What other point of view could they imagine? While Hebrews at the time would have been familiar with maps, and therefore could understand the concept of a bird's-eye view, there are no cues for such a viewpoint, such as, "Standing on a high mountain, I saw," or "The Lord looked down from heaven." In this context, they are told that the land was not always as they see it now. It was unformed and empty. Land and water were not clearly distinguishable. In fact, this is how modern science tells us the earth was at the beginning.

The picture, then, is like that of an empty stage, ready for the performance. God's Spirit "hovers" in anticipation. We are standing in the land of Israel, such as it was, waiting for something to happen.

My interpretation of Genesis 1 in terms of a stage metaphor draws from a famous scene in one of C. S. Lewis's *Narnia* books. Notice the beauty of the story as the observer (a boy named Digory) stands in one place, watching the land being filled:

> In the darkness something was happening at last. A voice had begun to sing. It was very far away and Digory found it hard to decide from what direction it was coming. Sometimes it seemed to come from all directions at once. Sometimes he almost thought it was coming out of the earth beneath them. Its lower notes were deep enough to be the voice of the earth itself. There were no words. There was hardly even a tune. But it was, beyond comparison, the most beautiful noise he had ever heard . . .
>
> Then two wonders happened at the same moment. One was that the voice was suddenly joined by other voices; more voices than you could possibly count. They were in harmony with it, but far higher up the scale: cold, tingling, silvery voices. The second wonder was that the blackness overhead, all at once, was blazing with stars. They didn't come out gently one by one, as they do on a summer evening. One moment there had been nothing but darkness; next moment a thousand, thousand points of light leaped out—single stars, constellations, and planets. . . . If you had seen and heard it, as Digory did, you would have felt quite certain that it was the stars themselves which were singing, and that it was the First Voice, the deep one, which had made them appear and made them sing. . . .
>
> The Voice on the earth was now louder and more triumphant; but the voices in the sky, after singing loudly with it for a time, began to get fainter. And now something else was happening.

Far away, and down near the horizon, the sky began to turn grey. A light wind, very fresh, began to stir. The sky, in that one place, grew slowly and steadily paler. You could see shapes of hills standing up dark against it. All the time the Voice went on singing. . .

The eastern sky changed from white to pink and from pink to gold. The Voice rose and rose, till all the air was shaking with it. And just as it swelled to the mightiest and most glorious sound it had yet produced, the sun arose . . .

The Lion was pacing to and fro about that empty land and singing his new song. It was softer and more lilting than the song by which he had called up the stars and the sun; a gentle, rippling music. And as he walked and sang the valley grew green with grass. It spread out from the Lion like a pool. It ran up the sides of the little hills like a wave. In a few minutes it was creeping up the lower slopes of the distant mountains, making that young world every moment softer. The light wind could now be heard ruffling the grass. Soon there were other things besides grass. The higher slopes grew dark with heather. Patches of rougher and more bristling green appeared in the valley. Digory did not know what they were until one began coming up quite close to him. It was a little, spiky thing that threw out dozens of arms and covered these arms with green and grew larger at the rate of about an inch every two seconds. There were dozens of these things all round him now. When they were nearly as tall as himself he saw what they were. "Trees!" he exclaimed . . .

The Lion was singing still. But now the song had once more changed. It was more like what we should call a tune, but it was also far wilder. It made you want to run and jump and climb . . .

Can you imagine a stretch of grassy land bubbling like water in a pot? For that is really the best description of what was happening. In all directions it was swelling into humps. They were of very different sizes, some no bigger than mole-hills, some as big as wheel-barrows, two the size of cottages. And the humps moved and swelled till they burst, and the crumbled earth poured out of them, and from each hump there came out an animal.³

Lewis's image of God singing things into existence is not foreign to Scripture. When God talks to Job, he uses the same image:

> Where were you when I laid the foundation of the earth? Tell me, if you have understanding. Who determined its measurements—surely you know! Or who stretched the line upon it?
> On what were its bases sunk, or who laid its cornerstone, when the morning stars sang together and all the sons of God shouted for joy?
>
> Job 38:4–7

Genesis 1 is a stage being set, ready for the drama of humanity. The setting of the stage is a beautiful song in its own right—the angels sing, and God gives commands to the inanimate land and sea. The observer is standing, like Digory, in one place—the land of the Hebrews—not moving variously about, hanging in space, looking down with a bird's-eye view, etc. Such perspectives would have been far from the mind of an ancient Hebrew.

Things Begin to Happen

> And God said, "Let there be light," and there was light. And God saw that the light was good. And God separated the light from the darkness. God called the light Day, and the darkness he called Night. And there was evening and there was morning, the first day.
> And God said, "Let there be an expanse in the midst of the waters, and let it separate the waters from the waters." And God made the expanse and separated the waters that were under the expanse from the waters that

were above the expanse. And it was so. And God called the expanse Heaven. And there was evening and there was morning, the second day.

And God said, "Let the waters under the heavens be gathered together into one place, and let the dry ground appear." And it was so. God called the dry ground Land, and the waters that were gathered together he called Seas. And God saw that it was good. And God said, "Let the land sprout vegetation, plants yielding seed, and fruit trees bearing fruit in which is their seed, each according to its kind, on the earth." And it was so. The land brought forth vegetation, plants yielding seed according to their own kinds, and trees bearing fruit in which is their seed, each according to its kind. And God saw that it was good. And there was evening and there was morning, the third day.

Genesis 1:3–13

Some day-age proponents equate "Let there be light" with the Big Bang. It is tempting, since the Big Bang is such a mysterious and cataclysmic event. But as I have noted, in Genesis 1:2 the land is already there. For a person standing on the land, in its formless state, the first event of significance is the ignition of the sun. Light appears, but—according to all we know from modern science—the earth was still cloudy, like the planet Venus. Light and darkness alternated as the earth turned on its axis, but the sun, moon, and stars were not visible in the sky. Everything was still murky and swampy.

Like Aslan singing in Narnia, God is directing, setting the stage. "Lights!" he commands, and the stage lights come on.

The next events also agree with what an observer might see. As the water cools, it liquefies, while previously it was hot and vaporous. There is now a clear horizon separating the sea below and the clouds above.

Several Hebrew scholars[4] have argued that the word "expanse" refers to a solid object; in other words, that the Hebrews viewed the sky as a solid dome, and the inspired text adopted this viewpoint. The only Scriptural evidence for this view is Proverbs 8:27–28: "When he established the heavens, I was there; when he drew a circle on the face of the deep, when he made firm the skies above, when he established the fountains of the deep." This passage in Proverbs is highly poetic, however, and it is not obvious at all that the "firmness" of the skies refers to a literal hard surface or just to their being well-established (recall that other poetic passages say that the earth cannot be moved, which all modern commentators take as metaphorical). The word "expanse" simply means a "surface." This can well refer to the surface of the sky as we see it. God, through the inspired writer, is not making a statement about the structure of the sky, but the appearance of it, as a shiny surface.

Next, the waters recede and the dry ground appears. Green plants begin to grow on the land. This, again, agrees with the basic picture of modern science.

Is it wrong-headed to think that God would give the ancient Hebrews a picture of things that modern science now tells us? Why not? The text is, after all, inspired. God is not concerned that they have every detail of geology, but to convey the basic concept that the formation of the land did not occur instantly, but took place over an extended period of time. This was not at all obvious to the ancients. As mentioned in chapter 6, Augustine felt that creation must have taken place instantaneously, and the extended period of six days must be allegorical.[5] Why should God, who can do anything, take so much time as six days? Genesis 1 tells us that indeed, God did not create everything instantaneously, but used a long process.

The "Day" and the "Evening and Morning"

Up to now I have not discussed what to many people is a key question of the day-age interpretation, namely, the meaning of the word "day" (*yom*, in Hebrew), and the meaning of the "evening and morning." As discussed in previous chapters, I do not think this is the main issue. Yet some commentators, such as John MacArthur, feel that these time words demand a literal twenty-four-hour-day view. In the above commentary, I have taken the days as indeterminate ages; in other words, as days from God's perspective as he sets the stage for the land of Israel. Is there any justification for this?

Perhaps it is no coincidence that immediately at the end of this passage, the word "day" is clearly used to refer to a period of more than one twenty-four-hour day. In Genesis 2:4, the singular form of the word "day" is used to refer to the whole period of the creation:

> These are the generations of the heavens and the land when they were created, in the day that the LORD God made the land and the heavens.
>
> Genesis 2:4

The word for day, *yom* in the original Hebrew, is used in many places in the Bible to indicate an indefinite period of time. Some examples of generic use of the word are 1 Samuel 8:8, 1 Samuel 8:18, and 2 Samuel 22:19, which have the sense of "in that day," that is, in that indefinite period in the past. The prophets frequently use the term "in that day" to refer to a future, indefinite period also called the "day of the Lord" (e.g., Isa. 2:20; 3:7; 3:18; 7:21 and many others). Perhaps the most dramatic example is in Zechariah:

> For behold, on the stone that I have set before Joshua, on a single stone with seven eyes, I will engrave its inscription, declares the LORD of hosts, and I will remove the iniquity of this land in a single day. In that day, declares the LORD of hosts, every one of you will invite his neighbor to come under his vine and under his fig tree.
>
> Zechariah 3:9–10

In verse 9, a "single day" is described in which God will remove the sin of the land. Most evangelical Christian commentators take this as referring, typologically, to the sacrifice of Christ, the cornerstone of the church. In the next sentence, "in that day" is used to refer to a time of peace and blessing. This clearly is not the same twenty-four-hour day as the day of Jesus's death on the cross, yet "that day" is identified with the "single day" in which the sin of the land is removed. The only way to read this is to view "that day" as an eschatological age ushered in by the sacrifice of Christ.

When the Hebrew writers wanted to make clear they meant a specific twenty-four-hour day, they used additional cues. For example, Genesis 7:11 literally reads, "In the six hundredth year of Noah's life, on the seventeenth day of the second month, on this day, all the springs of the great deep split open" (my translation). Moses goes to great lengths to specify that he means exactly this one day, and is not using the term generically.

As mentioned in chapter 5, another argument for the length of a "day" not being twenty-four hours is the classic statement of Peter:

> But do not overlook this one fact, beloved, that with the Lord one day is as a thousand years, and a thousand years as one day.
>
> 2 Peter 3:8

Peter, in turn, is referring to the Psalm of Moses, which says,

> Before the mountains were brought forth, or ever you had formed the earth and the world, from everlasting to everlasting you are God. You return man to dust and say, "Return, O children of man!"
> For a thousand years in your sight are but as yesterday when it is past, or as a watch in the night. You sweep them away as with a flood; they are like a dream, like grass that is renewed in the morning: in the morning it flourishes and is renewed; in the evening it fades and withers.
>
> Psalm 90:2–6

Notice that Moses, the author of Psalm 90, uses creation themes here: the formation of the earth, and the formation of Adam from the dust. In the same breath he says that God's days are not like our days.

Notice also that in Psalm 90 Moses uses the terms "morning" and "evening" in a generic sense—the grass flourishes in the "morning" and withers in the "evening." Often, young-earth commentators argue that the word "day" can refer to an indefinite period, but that when it is coupled with the term "evening and morning" it must mean twenty-four hours. Yet here we see Moses specifically using a "day" (yesterday) and "morning and evening" to refer to indefinite periods that are all passing away. The morning is the time period when humans flourish, and the evening is the time period when they fade. Grass does not literally spring up in the morning and wither on the same day; grass everywhere takes weeks to grow and die.

This use of morning and evening in a generic sense occurs elsewhere:

> Sing praises to the LORD, O you his saints,
> and give thanks to his holy name.
> For his anger is but for a moment,
> and his favor is for a lifetime.
> Weeping may tarry for the night,
> but joy comes with the morning.
>
> Psalm 30:4–5

> Like sheep they are appointed for Sheol;
> Death shall be their shepherd,
> and the upright shall rule over them in the morning.
> Their form shall be consumed in Sheol, with no place to dwell.
>
> Psalm 49:14

> The years of our life are seventy,
> or even by reason of strength eighty;
> yet their span is but toil and trouble;
> they are soon gone, and we fly away. . .
> Return, O LORD! How long?
> Have pity on your servants!
> Satisfy us in the morning with your steadfast love,
> that we may rejoice and be glad all our days.
>
> Psalm 90:10, 13–14

Each of these passages uses "the morning" to refer to a future time when suffering will be done away with. Other passages that use "the morning" to refer to a generic time period are Genesis 49:27, Job 4:20, Ecclesiastes 11:6, and Isaiah 21:12.

Some young-earth commentators have agreed that the "days" and the "morning and evening" can refer to indefinite time periods, but have insisted that the *numbering* of the days implies that they must be twenty-four-hour days. It is true that we can find no other passage in Scripture in which days are numbered and have a generic sense. But this uniqueness also makes it hard to insist on any particular interpretation—there is no other numbering of days in this form anywhere in Scripture, whether referring to twenty-four-hour days or any other time periods. The Hebrew form of numbering in Genesis 1 reads "and there was evening and morning, second day...and there was evening and morning, third day," without definite article or preposition (my translation). The only other place in Scripture where days are numbered at all is Numbers 29. The days there clearly refer to twenty-four-hour days. In that chapter, the formula is literally "on day the second . . . on day the third . . . ," quite different from the construction of Genesis 1. Not only that, after each day in Genesis 1, there is a "full stop" character in the Hebrew text, indicating a poetic or dramatic break, unlike the Numbers passage.

Genesis 1 is clearly different in style from the later historical chapters of Genesis and the historical books, because of its poetic structure. To say it is "poetic" does not mean it is not true; it simply has the form of a poem, like the Psalms, in which words are sometimes used symbolically and generically.

The Chorus Enters

> And God said, "Let there be lights in the expanse of the heavens to separate the day from the night. And let them be for signs and for seasons, and for days and years, and let them be lights in the expanse of the heavens to give light upon the land." And it was so. And God made the two great lights—the greater light to rule the day and the lesser light to rule the night—and the stars. And God set them in the expanse of the heavens to give light on the land, to rule over the day and over the night, and

to separate the light from the darkness. And God saw that it was good. And there was evening and there was morning, the fourth day.

And God said, "Let the waters swarm with swarms of living creatures, and let birds fly above the land across the expanse of the heavens." So God created the great sea creatures and every living creature that moves, with which the waters swarm, according to their kinds, and every winged bird according to its kind. And God saw that it was good. And God blessed them, saying, "Be fruitful and multiply and fill the waters in the seas, and let birds multiply on the land." And there was evening and there was morning, the fifth day.

And God said, "Let the land bring forth living creatures according to their kinds—livestock and creeping things and beasts of the land according to their kinds." And it was so. And God made the beasts of the land according to their kinds and the livestock according to their kinds, and everything that creeps on the ground according to its kind. And God saw that it was good.

Genesis 1:14–25

I have already mentioned in chapter 5 the problem of the sun appearing on Day 4, while the light appeared on Day 1. Another thing that has bothered people is that in this section, fish and birds appear before land creatures of all kinds. According to modern archaeology, birds appeared on the earth after at least some land creatures, which came after the fish.

These problems arise only if one has in view the entire earth, hanging in space, and if one takes the things that happen on these days as referring to global events. If one adopts the viewpoint of the land of Israel, they are much less problematic. In that case, one is looking at a stage being filled with performers: first the clouds part to reveal the sun, moon, and stars, then the flying population moves in, then the swimming population, and last, the land animals. All of them existed elsewhere, "backstage," and enter in order when called by God.

Why should the land animals enter last? Perhaps because the land was still not solid, a muddy marsh. The air and the water could be populated well before the land was truly solid. Or perhaps God simply kept land animals from migrating there until later, for dramatic reasons. We must not fall into thinking that everything must happen for natural causes independent of God's grand sense of drama.

The Hebrew verbs do not give the timing of the actual creation of sun and moon, or the creation of the land animals, in any specific way. Genesis 1:16 says, "And God made the two great lights," and Genesis 1:25 says, "And God made the beasts of the land." Hebrew does not have the same distinction between simple past ("made") and past perfect ("had made") tense that we have in English; the Bible does not give the exact timing of the event relative to other past events.

This is the same construction as Genesis 2:19, which in literal translation says, "And the LORD God formed from the ground every beast of the field and every bird of the heavens." Note that it says this *after* the creation of human beings is discussed. Liberal commentators take this as evidence that there are two competing creation stories side by side, as a contradiction of Genesis 1. Conservative commentators commonly take this line to mean only that God created the animals at some time. If it is legitimate to take the latter interpretation of Genesis 2:19, then it is just as legitimate to take the same interpretation of Genesis 1:16 and 1:25. Some actors who have been backstage are introduced on stage in an order that may not be the order they appeared in on the entire earth.

I have already discussed at length the tanninim, or "great sea creatures." The basic meaning of the Hebrew word is "great reptile monster." They are associated with the sea here, but that does not preclude that they could have been waders in the mud, that is, dinosaurs of all types.

Some commentators have made a great deal of the fact that at the beginning of Day 6, God uses an odd phrase: "Let the land bring forth living creatures" (Gen. 1:24). On previous days we mostly hear "God created" or "God made." Does this phrase refer to evolution, since the land "brings forth" the animals instead of God acting directly to create them? It is unlikely. First, why should evolution apply only to the land animals and not to sea creatures and birds? Second, the phrase "let the land bring forth living creatures" is written in the same style as the lines "Let the land sprout vegetation . . . and the land brought forth vegetation" in Genesis 1:11–12. The image is of the land bursting into fullness, not of the land acting as a causal force.

As discussed above, several day-age commentators have tried to identify the phrase "let there be light" with the Big Bang; so also many have tried to identify the creation of the "swarm" of the sea in Genesis 1:21 with the Cambrian explosion, a major, one might say miraculous, jump in the number of living species in a geologically short time. There may be a connection, but again, the focus of Genesis 1 is not on global events. It is also hard to see how the Cambrian explosion can be associated with the appearance of the birds and creatures of the air, since the fossils we see from the Cambrian explosion are all sea creatures.

The King Is Crowned

> Then God said, "Let us make man in our image, after our likeness. And let them have dominion over the fish of the sea and over the birds of the heavens and over the livestock and over all the land and over every creeping

thing that creeps on the land." So God created man in his own image, in the image of God he created him; male and female he created them.

And God blessed them. And God said to them, "Be fruitful and multiply and fill the land and subdue it and have dominion over the fish of the sea and over the birds of the heavens and over every living thing that moves on the land." And God said, "Behold, I have given you every plant yielding seed that is on the face of all the land, and every tree with seed in its fruit. You shall have them for food. And to every beast of the land and to every bird of the heavens and to everything that creeps on the land, everything that has the breath of life, I have given every green plant for food." And it was so.

And God saw everything that he had made, and behold, it was very good. And there was evening and there was morning, the sixth day.

Genesis 1:26–31

Day 6 is a long day compared to the others. After the introduction of the land animals, the same formula occurs that ended the other days: "And God saw that it was good." But on this last day, God continues. The creation of humankind is set apart from the creation of the land animals, although humans live on the land. Many commentators have discussed at great lengths both the concept of the image of God that is introduced here, as well as the mandate to fill the earth and to subdue it.

In the drama of creation, the human is introduced as king of creation. As discussed in chapter 3, a hierarchy is introduced: humans rule over the creatures of the land and the sea and the air, as well as the green plants; the animals rule over the green plants. The sense of this passage is not a *restriction*—do not eat animals—but a *commission* and a blessing: God says, in effect, "Go out and rule, for I have given them into your hand." Animals are not given a commission to rule over other animals, but they are given a commission to rule over the plants.

Using the word "land" here instead of "earth" does not diminish the authority given to humans; as in Genesis 1:1, the sense of the "land" here is "as far as the eye can see." Humans are commissioned to fill the land as far as they can go. The islands of Borneo are not in view, but this is not a restriction to stay in the Middle East; it is simply a restriction to what the ancient Hebrews knew. No one but God knew at the time how far people would go to fill the land.

The Sabbath

> Thus the heavens and the land were finished, and all the host of them. And on the seventh day God finished his work that he had done, and he rested on the seventh day from all his work that he had done. So God blessed the seventh day and made it holy, because on it God rested from all his work that he had done in creation.
>
> Genesis 2:1–3

I have already discussed this passage at length in chapter 5. The number seven is hallowed in Scripture as a sign of completion and perfection. There is no record of God picking up his tools again on an "eighth day." Such would be an incredible deflation of the story. God's work is done, and stays done.

The Introduction of the Characters

> These are the generations of the heavens and the land when they were created, in the day that the LORD God made the land and the heavens.
>
> Genesis 2:4

The form of this verse is used several times in the book of Genesis. It is the form of the beginning of a genealogical story. All the later occurrences refer to the generations, that is, offspring, of people. The next occurrence is in Genesis 5, after we have had the story of Adam and Eve:

> This is the book of the generations of Adam. When God created man, he made him in the likeness of God. Male and female he created them, and he blessed them and named them Man when they were created.
>
> Genesis 5:1–2

Note that this passage is directly parallel to the Genesis 1:27 statement about the creation of the human race. Many liberal commentators have argued that Genesis 1 and 2 give two different, contradicting creation accounts, presumably the result of two sets of editors working cross purposes to each other. Yet here we have a third creation account, which uses exactly the same formula from Genesis 1:27, and also the same format as Genesis 2:4—it ties the two accounts together.

When we look at other genealogy introductions, we discover that they all occur *after* their main characters have been introduced. Noah's genealogy begins after we have met him (in Genesis 5:29):

> These are the generations of Noah. Noah was a righteous man, blameless in his generation. Noah walked with God.
>
> Genesis 6:9

The same formula occurs in Genesis 10:1 (the sons of Noah); 11:10 (Shem, the ancestor of all "Semitic" peoples); 11:27 (Terah, the ancestor of Abraham); 25:12 and 25:19 (Ishmael and Isaac, respectively, Abraham's sons); 36:1 and 36:9 (Esau); and 37:2 (Jacob, the father of the nation of Israel).

The import of Genesis 2:4 is to introduce a genealogy story: the story of the descendents of the "heavens and the land." Adam is the son of the heavens and the land, so to speak, and his name, literally "red clay," affirms this. Genesis 2 is not a different account, but a narrowing of the account to events that directly concern the reader. In each of the generation accounts, the story is further narrowed, finally focusing just on the Israelites and their immediate neighbors.

The Garden

> When no bush of the field was yet in the land and no small plant of the field had yet sprung up—for the LORD God had not caused it to rain on the land, and there was no man to work the ground, and a mist was going up from the land and was watering the whole face of the ground—then the LORD God formed the man of dust from the ground and breathed into his nostrils the breath of life, and the man became a living creature.
>
> Genesis 2:5–7

Up to now in this chapter, I have used the English Standard Version translation of the Bible but substituted the word "land" for the word "earth" in each instance. But now, in chapter 2, verse 5, the English Standard Version suddenly switches to using the word "land" for the same Hebrew word. Why? One reason is that chapter 2 seems clearly to deal with a local story. Notice that this passage says that no rain came on the land, and there were no green plants. Conservative translators wish to reduce any sense of contradiction between Genesis 1 and 2; therefore, since in Genesis 1 we already have the green plants created, they switch

English words to indicate that no rain had come on *this* land.

The passage does not say that no rain had *ever* come on the land. At that time, the land was unfertile for two reasons. First, no rain had come, and second, there were no people to work the soil. The importance of rain places this story in the fertile crescent of the Middle East, where even in ancient times life depended on the rainfall, and lack of rain meant famine.[6] One of the main focuses of pagan religions was praying to their deities for rain, but here God is shown as lord of the rain, as well as creator of the human race.

Note that if the land had emerged from the waters just three days earlier (assuming that these events happen on Day 6, and the land appeared from under the waters on Day 3, in the young-earth view), then it hardly makes sense that the land would be dry and unfertile. For that matter, giving any discussion of causation for the lack of vegetation seems out of place, if the land had only just appeared. The sense of the text is that the land had been around a long time, so long that it had dried out. In this dry land, God creates a special garden, with water and a gardener.

> And the LORD God planted a garden in Eden, in the east, and there he put the man whom he had formed. And out of the ground the LORD God made to spring up every tree that is pleasant to the sight and good for food. The tree of life was in the midst of the garden, and the tree of the knowledge of good and evil.
>
> A river flowed out of Eden to water the garden, and there it divided and became four rivers. The name of the first is the Pishon. It is the one that flowed around the whole land of Havilah, where there is gold. And the gold of that land is good; bdellium and onyx stone are there. The name of the second river is the Gihon. It is the one that flowed around the whole land of Cush. And the name of the third river is the Tigris, which flows east of

Assyria. And the fourth river is the Euphrates. The LORD God took the man and put him in the garden of Eden to work it and keep it.

Genesis 2:8–15

This passage gives the sense that the Garden of Eden is not a mythical place in a "land far away," but a real place that you can find on a map. Moses goes out of his way to give us details relative to places the reader would know about: the land of Cush, the river Euphrates, and the products of nearby countries.

Geologist Caroline Hill makes a convincing case[7] for the location of the garden of Eden using modern geology. This evidence strongly supports the historicity of Genesis. Sadly, many conservative Bible scholars have not paid attention to her work because it rests on an old-earth view.

Figure 3 shows a map of the ancient Middle East from Caroline Hill's paper. Three of the four rivers that flowed through Eden still exist and are well known: the Euphrates, the Tigris (Hiddekel), and the Gihon (Karun). But one of the four rivers of the Garden of Eden has always been a mystery. Where is the Pishon and how does it get to the land of Havilah where there is much gold? Hill used satellite photos to show that underground is an ancient river bed that would have flowed several thousand years ago. This river cut across the Arabian peninsula to the mountains on the west side, where it is known that gold and onyx were found in ancient times. The conjunction of these four rivers occurs in the area where sat an ancient town called the city of "Eridu" (easily seen as a linguisitic variant of "Er-eden" or something similar), the oldest known city in the world, also known as a holy place to the ancient Mesopotamians. Note that Genesis 2:8 says "in Eden, in the east" as though this was a place people would recognize.

All good, so far. But this river lies *on top of* sedimentary geological layers that young-earth creationists would say were deposited in the flood of Noah. So do the Euphrates and Tigris rivers. To accept this convincing case for the historicity of Genesis, Bible scholars must either accept an old-earth view or believe that God created sedimentary rock at the beginning, before the fall and the flood.

Young-earth scholars must necessarily place the Garden of Eden in a long-lost land, since they believe sedimentary rock was created in the flood; therefore none of these rivers could have existed before the flood of Noah. For that matter, a flood large enough to create all sedimentary rock would wipe out all rivers. While Moses goes out of his way to place the Garden of Eden in our world, young-earth creationists make these geographical indicators irrelevant.

In summary, the opening chapters of Genesis have the feel of a drama unfolding, as first each element of the set is introduced, then the characters, and then the Garden, where the first act of the drama will play out. This view doesn't diminish our view of the historicity of the passage, it enhances it, as we see in many verses markers that locate the story in the real world, in the ancient Middle East. If we imagine ourselves on the "land" with an ancient Hebrew, listening to the story of the creation of the ancient hills and the great sea monsters, perhaps our sense of awe at the majesty of the story will increase.

Review Questions

1. Do you see any compelling textual reason why the word "earth" should be used in Genesis 1 instead of "land," even though ancient Hebrews had no concept of a globe? Does this word bias you to think of the entire globe?

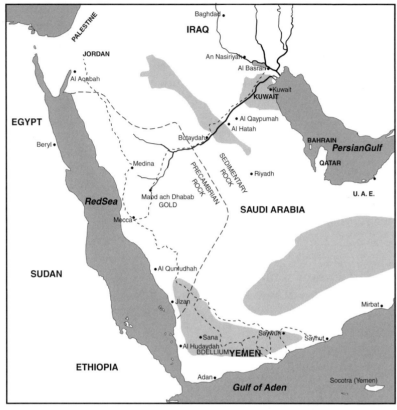

Figure 3. Map of the ancient Middle East. Three of the rivers mentioned in the Garden of Eden are well known—the Euphrates, the Tigris, and the Gihon (Karun). The fourth, the Pishon, can be equated with the Batin dry river bed, which runs across the Arabian peninsula, over top of sedimentary rock.

2. If a day must mean twenty-four hours, how do you understand the use of the word "day" in Genesis 2:4 to refer to the entire period of creation, or the prophetic use of "in that day," for example, Zechariah 3:10, to refer to an indefinite period?
3. Do you see a theater-stage aspect of the drama of Genesis 1? Does this help you to see the setting as a single point of view in the Middle East?

4. Why does Genesis 2 say that no plants had sprung up because there was no rain, if the land had emerged from the waters only three twenty-four-hour days earlier?
5. Does it excite you that the fourth river of the Garden of Eden may have been discovered, or does it bother you that this river lies on top of sedimentary rock? If the Tigris and Euphrates rivers lie on top of sedimentary rock, how could that rock have been created by the Flood of Noah when the Tigris and Euphrates existed at the time of Adam and Eve?

The Flood of Noah

8

So far, I have only discussed the creation story of Genesis 1 and 2, and I could leave it at that. On the issue of origins, however, one can not ignore the topic of the flood of Noah. The flood of Noah is used by young-earth creationists to explain all manner of geological data, in what is called "flood geology."

Since it is a stated goal of mine to reconcile the actual appearance of the earth with the teaching of the Bible, I ask, similar to the way I approached the study of Genesis 1, whether it is a *necessary* interpretation to read the story of Noah as a global, six-mile-deep flood that covered even Mount Everest, or whether it is *possible* to interpret this story as a more "local" flood.

I can already hear people saying, "Here we go down the slippery slope. First he wants to 'explain away' the creation week, now he wants to 'explain away' the flood, then what?" Is my goal to "explain away" all the miracles of the Bible, to find "natural causes" for them all?

Absolutely not. Just the opposite: I believe that all real miracles have "natural effects."

What do I mean by "natural effects"? Think for a moment of the story of the feeding of the five thousand. After this miracle, twelve baskets of leftovers were collected (John 6:13). The origin of the food was miraculous, but it left a definite, lasting physical record in the food that was picked up. The Bible text is very scientific in documenting the amount of food at the start, the number of people, and the amount of food at the end. The miracle had a "natural effect." If it had not, the people who were there might have wondered if they had imagined the whole thing.

In the same way, the man who was born blind and healed by Jesus had parents who could be interviewed (John 9:18). No one could provide a natural *cause* of the miracle, but they could readily discern natural, verifiable *effects* of it: "I was blind but now I see!" Miracles leave "tracks"—effects in the real world. This is what distinguishes real miracles from illusions and myths. Just as Paul was not afraid to produce lists of evidence for the resurrection (1 Cor. 15:5–8), we should not be afraid of looking at physical evidence.

In the case of small, local miracles, we are not surprised if no record remains of them to this day, since so many things can come along to erase the effects of a small event. But if a miracle was large, we expect that some tracks will remain. We expect that archaeology will be consistent with the records in the Bible of miraculous military victories, since the defeat of a major army will change history. If a miracle was *global*, then we expect global tracks. If the flood of Noah was truly worldwide, we expect to see worldwide evidence—the consequences of a thing like that are just not easy to erase.

Flood geologists, of course, claim to have such evidence. As I will discuss below, I find their evidence severely wanting. Despite many heroic attempts, there are simply too many gaping holes in their arguments.

Some people claim that the evidence of the flood is there, but that geologists are too unwilling to change their beliefs from traditional theories. Actually, in the past forty years, geologists have accepted two major, new theories in response to new discoveries. I have already mentioned in chapter 2 the "continental drift" theory. Forty years ago, this theory was considered quite controversial, but the weight of evidence (including the magnetic stripes on the ocean floor, discussed in chapter 2) forced geologists to accept it. More recently, a proposal that a large meteor hit the earth and wiped out the dinosaurs in a global cataclysm was scoffed at twenty years ago but has gained increasing acceptance as the evidence has accumulated. In each of these cases, revisions in the theories of the entire history of the world were supported by substantial, global evidence. Geologists accepted these new theories because they feel the evidence warranted a change of mind. No similar evidence has come up that has convinced geologists of a global flood.

God could, of course, have performed a global miracle and then studiously erased its tracks. This gets us back to the issue of appearances versus reality. Of course it is impossible to refute the idea that God did that, but then we are back to the question of whether we can trust anything we see in the natural world if God so effectively deceives us. The entire program of flood geology is to argue that science *as we know it* supports their interpretation of the Bible, not to propose unknown laws of physics that no one could guess at unless they were needed to cover up inconvenient facts.

Is it possible to find evidence for global miracles? Yes. I have already described how the Big Bang, the formation of the stars and planets, and the Cambrian explosion have no satisfactory explanation within modern science, so many concordantists consider these to be miracles.

While one might not agree that these events fit into a day-age model, these things show that it is possible to find global evidence of past major events. Many such discoveries have come in spite of fierce opposition from scientists who adhered to a more uniformitarian model.

The Scientific Case

Many people have written on the scientific problems with flood geology,[1] so here I will only summarize these difficulties. Again, the issue before us is primarily one of biblical interpretation, not science, but we do well to consider seriously the stakes involved.

To make the case, let me start from the opposite side and assume that the flood happened the way most people assume it did: a six-mile-deep flood covering the entire globe. God could certainly have done this in his infinite power. In order to have agreement with the text and our observations, however, this miracle would have required a whole series of miracles:

1. The miracle of transportation of millions of animals to the Ark from Australia, the Americas, Antarctica, and the islands. Some species of small animals that exist only in small niches in the ecology could never have made it to the Middle East without miraculous intervention.
2. The miracle of the compression of the animals in the Ark. The described volume of the Ark is not large enough for all of the millions of animal species plus the food and fresh water they would need for 150 days at sea.

3. The miracle of the feeding of the carnivorous animals on the Ark. If carnivorous animals came along, then many extra animals of other types had to come for food, unless meat was miraculously refrigerated.
4. The miracle of the caretaking of millions of animals by just eight people, including tons of dung production per day.
5. The miracle of the survival of the occupants of the Ark despite the huge heat production in a closed space. Rough calculations give temperatures of hundreds of degrees for so many millions of animals in a windowless box.
6. The miracle of the survival of special-climate animals (e.g. polar bears and penguins) on the Ark.
7. The miracle of the feeding of special diet animals (e.g. the koala) on the Ark.
8. The miracle of the creation of water out of nowhere and the destruction of that water afterwards. In a normal flood, water moves from one place to another, but in a global flood, new water would have to be created. Some people have proposed that the water resided in special holding tanks called the "springs of the great deep," and then went back there, but where are those holding tanks now, with enough water to cover the earth up to a six-mile depth? Alternatively, some have proposed that the water resided in a cloud "canopy," but to hold enough water to cover the earth to a six-mile depth, the canopy would have had to be so thick that the surface of the earth would have been hotter than Venus, and so dark that the moon and sun would not have been visible in the sky, in contradiction to Genesis 1:14.
9. The miracle of the non-sinking of the continents under the weight of that water. By normal laws of

physics, that much water would destroy the crust of the earth.
10. The miracle of the survival of the trees and plants underwater. Plants normally suffocate after just a short time underwater. Yet in Genesis 8:11 a dove returns with a fresh olive leaf shortly after the waters subside. The tree could have grown miraculously quickly from a seed after the flood, but the text does not indicate this.
11. The miracle of the survival of fresh water fish in salt water (or, salt water fish in fresh water). Unless God miraculously kept the water from mixing, half of the species of fish would have died.
12. The miracle of the survival of amphibious and tidal pool creatures on the ark. Certain species need particular conditions of water environment. In addition to bringing food and water to drink for the regular animals, Noah would have had to set up special aquariums that changed the type of water at different times of day.
13. The miracles of the survival of worms, insects, etc. underwater (or, the miracle of the transportation of all insects, worms, etc. to the Ark, including all the necessary ingredients for termite hills, ant colonies, etc.). If every type of worm and bug had to come on the ark, it would have been very exciting, indeed.
14. The miracle of the sorting of fossils under water into layers. I have already mentioned in chapter 2 that fossils are sorted by type into layers. This sorting is not by weight, but by land versus sea attributes. Land animals lie on top of layers of limestone (sea fossils), etc.
15. The miracle of the upright trees under water. A coal mining engineer from near where I live in Pennsylvania pointed out that they typically find vertical tree trunks in coal seams. These are not all at one

level, as one might expect for a suddenly flooded forest, but at many different levels. This seriously challenged the faith of this Christian man with a young-earth background.

The problem with the above list is not that God could not have done all of it, but that there is no indication in the Bible text of any of it. People must "read in" to the text all of the above miracles. If we are to be faithful to normal rules of interpretation, can we in good conscience read in so many things?

Some advocates of a global flood have argued that the above miracles were not necessary because the earth was relatively flat and the continents were not divided before the flood, and there were relatively few animal species, which then mutated and adapted to the new environments created after the flood. In so saying, they not only affirm evolution, which I reject, but they make the forces of evolution occur a million times more efficiently than any Darwinist would claim! Mountains spring up in a matter of weeks, continents tear apart and skid across the oceans at high speeds, nuclear radiation pours in from the skies creating mutations that lead to whole new species of plants and animals that are adapted to special environments like tidal pools or mountain tops, all within the few hundred years between Noah and Abraham. As one young-earth advocate said to me, no wonder Noah got drunk!

Again, of course God could do this miraculously, but where is the textual evidence? Flood geologists point to the phrase in Genesis 10:25, that in Peleg's day "the land was divided," but this is an awful lot to read into one phrase. There are much plainer ways to read that verse: either that an earthquake occurred, or even that the land was "divided" into different countries by social agreement, in the same way that Lot and Abraham divided the land. Furthermore, as I began my discussion in chapter 3, any normal reader would assume that the world of Genesis 1 is *our world*, and that the animals in it are the same as those we have. In the flood geology viewpoint, God essentially re-creates the entire world after the flood of Noah. Just as young-earth advocates read in an entire re-creation of the world in the curse of Genesis 3, flood geology reads in another entire re-creation, not mentioned in Scripture, after the flood. Flood geology has God creating the world not once, but three times, without any direct Scriptural evidence, and in contradiction to the statement in Genesis 2:1 that the entire creation was *completed* by the seventh day.

"The Waters Flooded the Land"

Why do we need to posit all those miracles? Mainly because people take Genesis 7:19–20, which says that the water covered all the mountains, as including Mount Everest and every other mountain range on earth. Looking at those verses, however, a close reading already creates problems for those who believe in a global flood. The literal reading is "the water rose twenty feet, and all the high hills were covered" (the New International Version gives a similar translation in a footnote). The passage says that the water rose only twenty feet, not six miles. For no reason other than to make sure there would be enough water for a global flood, this verse is frequently altered to "the water rose to twenty feet higher than the highest mountains." This latter reading is *not* the "literal" reading; it is interpolated, that is, read in to the text. Some people might argue that none of the world's mountains were higher than twenty feet until after the flood, but this leads to the outrageously fast form of evolution after the flood that I discussed above.

The need to rewrite Genesis 7:20 follows from another interpolated translation. As discussed in chapter 7, the word translated "earth" everywhere in Genesis 7–9 is most directly translated "land," with no more or less specificity than that word has in English—the same word as in "all the land came to Egypt" (Gen. 41:57). The translation of this word as "earth" or "world" biases the reader to understand this as the "globe" or "planet," but this meaning is not in the original text. Did everyone on the entire globe come to Egypt? Likewise, the most natural meaning, in Moses's day, of "the water covered all the land," is "the water covered everything as far as the eye could see" and not "Mount Everest, on the other side of the globe, which you don't even know about, was covered."

Note the different feel of the passage if we make this change. The following is Genesis 6–7, from the English Standard Version, with *adamah* translated as "ground" and *eretz* translated as "land."

> When man began to multiply on the face of the ground and daughters were born to them, the sons of God saw that the daughters of man were attractive. And they took as their wives any they chose. Then the Lord said, "My Spirit shall not abide in man forever, for he is flesh: his days shall be 120 years." The Nephilim were in the land in those days, and also afterward, when the sons of God came in to the daughters of man and they bore children to them. These were the mighty men who were of old, the men of renown.
>
> The Lord saw that the wickedness of man was great in the land, and that every intention of the thoughts of his heart was only evil continually. And the Lord was sorry that he had made man in the land, and it grieved him to his heart. So the Lord said, "I will blot out man whom I have created from the face of the ground, man and animals and creeping things and birds of the heavens,

for I am sorry that I have made them." But Noah found favor in the eyes of the LORD.

These are the generations of Noah. Noah was a righteous man, blameless in his generation. Noah walked with God. And Noah had three sons, Shem, Ham, and Japheth.

Now the land was corrupt in God's sight, and the land was filled with violence. And God saw the land, and behold, it was corrupt, for all flesh had corrupted their way in the land. And God said to Noah, "I have determined to make an end of all flesh, for the land is filled with violence through them. Behold, I will destroy them with the land."

Genesis 6:1–13

For behold, I will bring a flood of waters upon the land to destroy all flesh in which is the breath of life under heaven. Everything that is on the land shall die. But I will establish my covenant with you, and you shall come into the ark, you, your sons, your wife, and your sons' wives with you.

And of every living thing of all flesh, you shall bring two of every sort into the ark to keep them alive with you. They shall be male and female. Of the birds according to their kinds, and of the animals according to their kinds, of every creeping thing of the ground, according to its kind, two of every sort shall come in to you to keep them alive. Also take with you every sort of food that is eaten, and store it up. It shall serve as food for you and for them." Noah did this; he did all that God commanded him.

Then the LORD said to Noah, "Go into the ark, you and all your household, for I have seen that you are righteous before me in this generation. Take with you seven pairs of all clean animals, the male and his mate, and a pair of the animals that are not clean, the male and his mate, and seven pairs of the birds of the heavens also, male and female, to keep their offspring alive on the face of all the land.

For in seven days I will send rain on the land forty days and forty nights, and every living thing that I have made I will blot out from the face of the ground." And Noah did all that the LORD had commanded him.

Noah was six hundred years old when the flood of waters came upon the land. And Noah and his sons and his wife and his sons' wives with him went into the ark to escape the waters of the flood. Of clean animals, and of animals that are not clean, and of birds, and of everything that creeps on the ground, two and two, male and female, went into the ark with Noah, as God had commanded Noah. And after seven days the waters of the flood came upon the land.

In the six hundredth year of Noah's life, in the second month, on the seventeenth day of the month, on that day all the fountains of the great deep burst forth, and the windows of the heavens were opened. And rain fell upon the land forty days and forty nights.

Genesis 6:17–7:12

The flood continued forty days on the land. The waters increased and bore up the ark, and it rose high above the land. The waters prevailed and increased greatly on the land, and the ark floated on the face of the waters. And the waters prevailed greatly on the land and covered all the high hills under the whole heavens. The waters rose fifteen cubits, and covered the hills.

And all flesh died that moved on the land; birds, livestock, beasts, all swarming creatures that swarm on the land, and all mankind. Everything on the dry land in whose nostrils was the breath of life died. He blotted out every living thing that was on the face of the ground, man and animals and creeping things and birds of the heavens. They were blotted out from the land. Only Noah was left, and those who were with him in the ark. And the waters prevailed on the land 150 days.[2]

Genesis 7:17–24

One can easily read this as a relatively local flood that has the purpose of killing all humans—not every rabbit on the top of the Alps. Some may object to this, but the word "land" could not possibly have meant a globe to people who knew nothing of the shape, let alone the size and scope, of the earth. "Land" means "the place where people live," no more and no less. The size of the boat would be about right for the number of animals in the Middle East, which God protected because he did not want any part of his creation to be utterly destroyed.

One of the objections people have to a local flood is that they assume it means that some people escaped the flood because they were living elsewhere. This does not follow. Scripture indicates that the flood killed all people in existence except for Noah and seven others (2 Peter 2:5), that is, those who were on the Ark. Some may wonder how a localized flood could accomplish this. As the story of the tower of Babel in Genesis 11 makes clear, people even after Noah resisted the command of God to "fill the earth" and remained clumped together in the Middle East until God dispersed them. It is not hard to suppose that, before Noah, they did the same. Therefore, a flood restricted to the Middle East would kill every human. If all people descended from Noah, the oral tradition of a great flood, which is so well documented among various tribes around the world, would go with them as they spread to fill the earth after Babel.

The New Testament references to this passage also do not specify the size of the flood. Peter says that the *kosmos* was destroyed (2 Peter 3:6), a word that typically refers to the political order (hence, "cosmopolitan"). This same word is used in Acts 17:6, "these men who have turned the world upside down," which very few people take as referring to the entire globe. In Luke 17:26–27 (parallel with Matt. 24:37–39), Jesus says that the flood "destroyed them all," which again puts the focus on the people, not the destruction of the top of the Alps. In Luke 17:28–29, Jesus draws a parallel with Sodom, which was clearly a local destruction. Both Luke 17:27 and 17:29 literally say that "all" ("panta," in Greek) was destroyed, yet we know that in the case of Sodom this does not mean everything in the world, but simply all the people (and their "world," or culture.) First Peter 3:20, the only other mention of the flood of Noah, also discusses only the people and not geography.

The only statements in the story of Noah that give difficulty for the local flood view are God's promises, "Never again will I curse the ground because of man," (Gen. 8:21) and "Never again shall there be a flood to destroy the land" (Gen. 9:11). If this was a local flood, then did God mean there will never be any more local floods or other natural disasters? Clearly not, since we have them all the time; even in Moses's day, the Nile and the Euphrates flooded all the time. Did he mean, "Never again will the land of Israel have a flood?" That seems *too* local. God's statements appear to have sweeping implications, not restricted to application to Israel (although the land of Israel is frequently called simply "the land" in Scripture).

A key to understanding what God meant is found in the next verse, Genesis 8:22, which I have already discussed in chapter 4 as a key "balance" passage:

> While the land remains, seedtime and harvest, cold and heat, summer and winter, day and night, shall not cease.

God says, "I will *not* do what I have done, but *instead*, there will be balance in nature." Recall that the relationship between the sea and the land is another essential "balance" in Scripture. Moreover, it was highly important to the Israelites that the "boundaries of the sea" were set by God as part of the balance of nature (Job 38:8–11; Ps. 104:6–9; Jer. 5:22)—that dangerous forces like the sea and night could go only so far, and no farther. The essence of God's promise in Genesis 8:21 is that the balance of nature would not be overturned again. The flood was big enough to unbalance all of nature: the sea did not stay within its confines, and also, apparently, the seasons were altered, leading to a forty-day rain, among other things. Therefore, while one need not posit an immersion of Mount Everest, the flood in some very real way upset the natural balance of the entire earth and was a curse on all of life due to humanity's sin.

The same thought follows Genesis 9:11. In Genesis 9:15, the promise is restated as, "And the waters shall never again become a flood to destroy all flesh." An unbalancing catastrophe that kills all life, in particular all people, will not be repeated.

I do not plan to make any proposals on how the flood happened. Various proposals have been made from melting of ice caps, breaking of a continental divide, the end of the Ice Age, a meteor impact in the sea, etc. It was clearly miraculous in many ways. I do not take the view that it is merely a fanciful parable, since it is plainly viewed as historical by Peter and Jesus and others in the Bible. But the text does not demand us to "globalize" the story, all the colorful art that has been created for children notwithstanding.

As an aside, I note that Genesis 8:21, read literally, can be taken to say that the curse on the ground of Genesis 3:17 has been lifted. In exactly parallel language to Genesis 3:17, in which God curses the ground on account of Adam and Eve, here he lifts the curse in a clear parallel:

> Cursed is the ground because of you (Gen. 3:17)

> I will never again curse the ground because of man (Gen. 8:21)

Most commentators, of course, would argue that the second text refers to a different, "temporary" curse that is lifted, not the original curse. But the text does not say that. Clearly, we as humans are still under the curse of alienation from God, but one could argue from this that God has ceased to curse animals or upset the natural balance, on account of us. If this is true, of course, then the argument that the curse on the land of Genesis 3:17 implies numerous changes of animals from herbivorous to carnivorous varieties, etc., cannot be correct unless God changed them all back to their original versions after the flood.

I don't believe this argument, but in my opinion it has the same level of merit as the argument for animal vegetarianism before the fall, which I discussed in chapter 3. That argument insists that the parallelism of the form of the edicts of God in Genesis 1:29 and Genesis 9:3 requires us to take the second edict as a deliberate undoing of the previous edict. By that standard, the parallelism of Genesis 3:17 and Genesis 8:21 must be taken in the same light.

The Flood as a Type

Just as the Garden was an archetype of heaven, but not the same as the real thing, so the flood was an archetype of God's judgment:

> Just as it was in the days of Noah, so will it be in the days of the Son of Man. They were eating and drinking and marrying and being given in marriage, until the day when Noah entered the ark, and the flood came and destroyed them all.
>
> Luke 17:26–27

> For they deliberately overlook this fact, that the heavens existed long ago, and the earth was formed out of water

and through water by the word of God, and that by means of these the world that then existed was deluged with water and perished. But by the same word the heavens and earth that now exist are stored up for fire, being kept until the day of judgment and destruction of the ungodly.

2 Peter 3:5–7

Just as the flood came unexpectedly, so will the final judgment. Just as the flood upset the balance of nature and unleashed God's raw power, so will the final judgment. The story of the flood of Noah is not primarily about cute animals from all over the world. It is a warning of God's sudden wrath on sinners, and a reminder that the balance of nature cannot be taken for granted, but is a specific providence of God.

Both the creation story of Genesis 1 and the story of Noah have been sanitized for the modern mind. We saw in chapter 5 that the creation story of Genesis 1 contains elements (the darkness, the sea, the leviathan) that would have indicated to a Hebrew reader God's control over dangerous and powerful forces, a message echoed in various creation passages throughout the Bible, such as the end of the book of Job, Psalm 104, and Isaiah 45. This was in direct contrast to other creation stories of the time, which presented dark forces like the leviathan as ungovernable. Our Sunday school version presents Genesis 1 as an idyllic English garden with nice, non-carnivorous animals. In the same way, the Sunday school version of the story of Noah puts the emphasis on the amazing miracle of all the different kinds of animals and takes it off the main emphasis—the sudden wrath of God which can strike without notice. If we remove that part of the story, and say that only animals from the Middle East went on the boat, I can imagine some people will ask, "Then what is the point?" If we remove the animals from all over the world, true, we don't have a nice children's story any more, but we do have a story, like the story of Sodom and Gomorrah, of God's powerful wrath and destruction.

Again, there is no doubt about God's ability to cause a global flood. The issue is primarily one of biblical interpretation—do the passages demand such an interpretation, or have we been overly influenced by tradition? The scientific data cause us to take a second look at the traditional interpretation, because things appear inconsistent with flood geology. If we are to conclude that the Bible cannot be interpreted any other way, then, as in the case of the age of the earth, if we are to be honest with the scientific data we must simply adopt a view of "apparent history" in which evidences of the flood have been miraculously hidden.

I want to stress that I view the idea of "apparent history" in regard to the flood as intellectually viable, just as I view the idea of "apparent age" as viable in regard to the creation. On one hand, it grates against us to distinguish between "essences" and "appearances" as the Roman Catholic Church did in the Middle Ages. On the other hand, a belief in "things unseen" is essential in the Christian faith.

The earth appears to be able to last for millions of years, but on faith we believe that Christ could return and destroy it at any moment. One can argue that just as we believe that the earth is "reserved for fire" (2 Peter 3:7 NIV) even though "everything goes on as it has since the beginning of creation" (2 Peter 3:4 NIV), so also we believe, despite all appearances, that the earth was created in waters and "by these waters also the world of that time was deluged and destroyed" (2 Peter 3:6 NIV).

As mentioned above, the flood is an archetype for the final judgment, and miraculous "hiddenness" may be part of that parallel. If I were convinced that my Bible interpretation were wrong, I would adopt this view. One thing I could not do, without being utterly dishonest in regard to my scientific experience, would be to adopt the view of Henry Morris and some other flood geologists, that science tells us that the earth *appears* to have had a global, six-mile-deep flood. It does not.

My sense of the analogy in 2 Peter 3, however, is not that Peter is saying that we must accept both the flood and the second coming on blind faith. Rather, he seems to say that the historicity of the flood, including its unforeseen suddenness, should serve as a reminder to us so that our faith in the future is not blind faith. He therefore acknowledges an asymmetry between the past and the future, putting things in the past into the category of things seen.

Review Questions

1. Is it legitimate to change the meaning of Genesis 7:20 from "the water rose twenty feet, and covered the high hills," to "the water rose twenty feet above the highest mountains"?
2. If the word "land" were used in English translations of Genesis 7–9 instead of the word "earth," as is legitimate from the original text, would this affect the way you read this passage?
3. If the world was utterly re-created after the flood, with thousands of new species, mountains, and breaking apart of the continents, why doesn't the Bible say anything about this?

Implications for Theology

9

A change in interpretation of the Bible always has implications for theology. When we adopt a new model it means we must change those presuppositions that made us resist the new interpretation in the first place. For example, many of us are familiar with the sweeping changes in theology implied by changing our answer to the question, "Is the church now the new Israel, or do the promises of the Old Testament apply only to physical Jews?" Answering that the church is the new Israel leads to "covenantal" theology, which sees the church as God's triumphant kingdom, while rejecting this equation leads to "dispensational" theology, which views the present church age as a "parenthesis" in God's plan of history which centers on the Jews. In the same way, changing to a view of the earth as very old will have many important consequences.

How Do We Look at Nature?

Deciding whether a world with animal death is "very good" or "very bad" will have great implications for how we see the world around us. The young-earth view, which sees all animal death as "very bad," sees our world as essentially twisted and wrong, as utterly different from the "very good" world that God made, while I have argued that the message of Genesis 1 is that the powerful forces of the world in which we actually live are under God's control and give glory to him. The latter view implies several specific points:

- *The forces of nature are not "evil," even powerful forces that threaten us.* We tend to personify nature, but thunderstorms and animals do not sin. They become "evil" only in relation to us, when God uses them as forces of judgment against us.
- *The lives of animals are "vanity," fleeting and forgotten, but God still values them and rejoices in them.* God is proud of his handiwork, including carnivorous animals like the lion and the leviathan, glorying in them in passages like Job 38–41; Psalm 104; and Genesis 1. Like the grass, the animals pass away, perhaps for millions of years, without a human witness, but nevertheless give glory to God.
- *Humans are unique.* Only human beings were exempted from death via the special Tree of Life. Humans have a unique eternal destiny to rule over not only nature, but angels. Their lives on the earth in the "probational" state of the covenant of law "for a little while" was not meant as their ultimate destiny, but they were intended to "grow up" to rule over angels, presumably after they had fulfilled the command to fill the earth. The fact that we now share the same fears and concerns as the animals, as part

of the curse, does not take away the fact that we were meant for an eternal destiny.

- *Although human beings are the highest creation of God, the entire universe does not revolve around us, but around God.* It is perhaps flattering to think that our sin caused the second law of thermodynamics to exist, or all manner of animals such as sharks and vultures to spring into being. It is more humbling to think that perhaps our sin had no more effect than that we were ignobly flung out into the wilderness of an existing world that continued glorifying God as it always had. That that world incidentally ground us underfoot as God judged us by treating us the same as the transient animals and plants.

- *There is beauty in the balance of nature.* If we embrace the view that we should rejoice in the balance of nature, including the balance of predator and prey, some worry that this will lead us to accept human death in the same way, as part of the "cycle of life," while the Bible presents death as an enemy to be defeated (1 Cor. 15:26) with whom we must not make peace (Isa. 28:15–18). My response to this is simply that we must never allow people to equate humans with animals, that we must insist that humans have a unique soul and destiny and that we were not meant to live like the animals. If we teach that animal death is a horrible evil, we effectively affirm the vegetarian, New Age teaching that animals have spirits with eternal significance.

Some have argued that every animal death is an "object lesson" for us of the sin of Adam and Eve, and that every time we see an animal die, we should lament how we have twisted the universe so much. Yet if this object lesson is so important, why is it never taught in Scripture? Scripture teaches us two things about carnivorous animals: that God is glorified by them, and that they are used to judge us. Nowhere does Scripture teach us that carnivorous animals are woefully twisted from their earlier, better, herbivorous form.

One might ask, if God is ultimately going to create a world with no darkness, no sea, and no carnivorous animals in the new creation, then why would he make a world that included those things in the first creation? If such things will not exist in heaven, how could he be glorified by them in the world of Genesis 1? This is essentially the same as the question of why God did not create humanity in a state of grace in the first place, instead of in the probational status of the covenant of law. God's only answer is given in Isaiah 45—who are we, the pot, to question the Potter? The fact is that God did not do it that way. The Bible teaches very clearly that the world of Genesis 1–2 was not like heaven. With elements like the darkness, the sea, and the leviathan, not to mention the forces of judgment and the law, the status of humanity in Genesis 1–2 was not like their ultimate state in eternity. This brings up another implication:

- *God is good, but he is not "nice."* As C. S. Lewis said about the lion Aslan, he is good but he is not "safe." Perhaps this is the hardest lesson of all. We believe that God should be the way we want—focused on making us happy and making things that we like. God instead reveals in creation his terrible power and potential for wrath.

Where Are We Going?

As discussed in chapter 3, a very basic question involved in this debate is where we think we are going, at the end of all history. A common view says that we go back to the garden of Genesis 2, that heaven is a restoration of the paradise lost of Adam and Eve. I have argued, in contrast, that heaven is far more. Adam and Eve did not lose heaven, they were never in it. Instead God placed them in a special, temporary place with a special mandate to fill the earth by marriage, childbearing, and creating culture, and with clear dangers both physically (from carnivorous animals like the leviathan) and spiritually (from Satan in fleshly form as a snake).

This gets to the basic issue of how we view all of culture. The "garden" view of heaven fits naturally with the common feeling of Christians that cities are evil, and that isolation from crowds and separation from mainstream culture are the ideal. The Bible does not present such a view. The final state in heaven is a *city*—the New Jerusalem of Revelation 21, which we see prefigured throughout the Bible. As Augustine wrote centuries ago, the conflict in the Bible is not between city and country, but between two cities, the City of God (prefigured by Jerusalem) and the City of Man (prefigured by Babylon). We look forward to the great city culture of God's kingdom, we do not look back to an idyllic, pastoral garden with no animals except fuzzy rabbits and house pets.

Our world always has been, even before Adam and Eve fell, a wild place full of powerful, God-glorifying forces meant to be tamed by the king of creation, namely, the race of Adam, and woven by us into the fabric of a greater, everlasting culture of the city of God. Our sin has made this job far harder, but the focus is the same: taming the wilds as God's caretakers of the earth.

Appearances vs. Reality

Many people have debated the question of to what degree the appearance of the natural world should agree with our theology. A complete discussion of this issue is beyond the scope of this work; I have previously argued[1] that science and theology cannot exist in two separate worlds; Francis Schaeffer has written brilliantly along similar lines.

My entire approach has assumed that there should be agreement, and that experience, including science, may legitimately affect our Bible interpretation. Let me summarize again several things I am *not* saying. I am not saying that science "trumps" the Bible, or that science must have the final word. I have argued that the two should agree, and while in many cases this means that science may force us to take a second look at the Bible, as in the case of the theory of the moving earth, which Galileo promoted to the dismay of the Roman Catholic Church, it also means that the Bible may force us to go back to take a second look at science. The concordantist interpretation of Genesis 1 has several implications for science. The concordantist model may be wrong, but it is viable enough to cause us to go to nature and look for the tracks the miracles of the creation story may have left in the real world.

I also am not saying that there must be a natural cause for every miracle. I am simply saying that real miracles have effects, and if a person proposes that a certain miracle was global, it is reasonable to look for consistent global evidence. Lack of such evidence should weaken the claim that the miracle was in fact global.

Finally, I am also not saying that there is no place for faith in God despite all appearances. Our faith in the second coming is based entirely on his Word—the universe looks like it could easily last millions of years. As I argued in chapter 2, however, we must treat the past and the future differently. We take a step of pure faith into the future precisely because we have the knowledge of the consistency of God in the past. It is blind faith, not sound faith, to ignore the past and to ignore issues of consistency. "I know whom I have believed, and I am convinced that he is able to guard until that Day what has been entrusted to me" (2 Tim. 1:12).

Should We Read in the "Grandest Possible Miracle" for God?

A friend of mine told me that if he adopted my view of the flood of Noah, it would take away the majesty of the passage—it would no longer be the grand, global miracle he had learned from Sunday school, and this would diminish the glory of God in the passage. Others may feel the same way, that my reading of the stories of the curse and the flood makes them less dramatic, less "radical," because they do not involve sweeping, universal changes in the entire order of the world, but instead focus on humanity and God's judgment. That may disappoint us, but it is hardly a good principle of Bible interpretation that we should always look for the most grand, radical interpretation of a passage. We must take the interpretation that most naturally fits the passage in the context of the rest of Scripture and of legitimate extra-biblical knowledge. Whether or not that interpretation fits our idea of what is "radical enough" is neither here nor there. Jesus not only walked on water, he also cursed a small fig tree. Should that bother us because it is not grand enough?

I feel that the common interpretations of the flood and the creation in terms of grand, global miracles has taken the focus off the real importance of these stories and onto "grand magic." When school children learn about the flood of Noah, they focus on all the zebras and koalas or unicorns that supposedly made the trek from far-off regions of the globe, and they neglect the real significance of the story: that if God were to stop the balance of nature even for a short time, humanity would be swept away in an instant, and that God's ultimate judgment will come without warning in the same way. The standard interpretation of the creation story also leads people down the wrong path. Instead of taking from the story the teaching that all of the threatening and dangerous forces in our world are in balance under the sovereignty of God, school children get the idea that God is only glorified by what is idyllic and peaceful, and that therefore something must be terribly wrong if dangerous forces exist.

Young-Earth Creationism Is Incompatible with the Intelligent Design Movement

A few years ago, I participated in an Internet forum on intelligent design in which numerous young-earth creationists also participated. I found that they consistently undermined the arguments of the intelligent design movement. This is not accidental, but intrinsic to the premises of young-earth creationism.

- *Intelligent design says that we see good design in nature; young-earth creationism sees badness and twistedness in nature.*

In the young-earth view, all carnivores, parasites, and birds and animals that eat dead meat are evidence of badness, not goodness. Young-earth creationists see a lot of good in nature, of course. But they pick and choose what is good and what is bad, so they have no argument against the evolutionist who says that a particular example is bad design. Evolutionists commonly argue that things must have evolved because they are poorly designed, and therefore an intelligent designer could not have been involved,[2] for example the panda's thumb[3] or the human eye.[4] Someone in the intelligent design movement might respond that in fact, when the details are understood, we can see that it is very well designed; for example, Jerry Bergman has argued convincingly that the human eye is very well designed,[5] and a Japanese study found the panda's thumb to be much more subtly designed than first thought.[6] But the young-earth creationist essentially agrees with the evolutionist that many things we see are badly done, they are things "twisted" or "fallen." If young-earth creationists say that parasites or carnivores are twisted and misshapen forms, then why not the human eye or the panda's thumb?

The intelligent design argument says that we see mostly good design. Of course there is dysfunction, but systems of repair, regeneration, compensation, immunity, and protection against dysfunction show even more aspects of good design. We see good design not only in single life forms, but also in the general balance between carnivore and herbivore, blossoming and decay, etc.

Some young-earth creationists would say that carnivores and parasites are good design, but that they were created by God after the fall as a curse. But if they are good, then why not allow that they could have existed before the fall? As discussed in chapter 3, the only evil in their existence is that people with eternal souls should be subject to these dangerous forces. By themselves, these creatures are not evil, whether before or after the fall.

- *Young-earth creationists often ascribe much greater evolutionary power to natural forces than Darwinists do, since they say that all the diversity of life appeared naturally after the fall or the flood.*

The Bible texts do not mention dramatic new creation events after the fall or the flood. Therefore, young-earth creationists typically say that the changes from non-carnivorous to carnivorous or nonparasitic to parasitic at the fall were minor. In the same way, to explain how certain types of life exist that are perfectly adapted to special environments that could only have existed after a global flood, or to account for the great number of species that could not have fit on the Ark, young-earth creationists typically argue that animal species rapidly diversified to adapt to new environments after the flood.

In saying this, young-earth creationists actually make evolutionary forces much greater than even the most ardent evolutionist would vouch. A change from a herbivorous to a carnivorous animal requires only a small change! The appearance of a new species perfectly adapted to tidal pools takes only a few years!

Alternatively, some young-earth creationists will say that God created new species miraculously after the fall and the flood. In so saying, they do great violence to the texts, introducing three creation stories where the Bible appears to have only one.

- *Young-earth creationists argue that we should not "go where the evidence leads," but that it is legitimate to hold out for a favored theory despite all evidence.*

In embracing the "appearance of age" hypothesis, young-earth creationists say that scientific arguments should not, after all, be settled on the basis of observations.

The intelligent design movement presents evidence that Darwinian mechanisms cannot account for the diversity of life we see, and tries to define attributes of design, which we also see. Intelligent design proponents frequently fault evolutionists for closing their eyes to the evidence, and for forcing an interpretation despite all appearances to the contrary. But the evolutionist can respond, "It just *appears* to be designed. I am taking a position no different from your approach to the age of the earth."

I do not mean that no one should ever hold off from embracing new evidence. Science has many examples of new evidence that is loudly proclaimed as confirming a theory, only to fade away when examined closely by others. But when faced with an overwhelming preponderance of firm evidence, it becomes irrational to hold out. If the Pharisees had produced the dead body of Jesus and paraded it in the streets, it would have been foolish to be a Christian. If we seal ourselves off from this risky aspect of our faith, that it can be falsified, then we make *any* theory anyone wants to believe just as good as our beliefs.

- *Young-earth creationists undermine the evidence for fine-tuning in the design of the universe.*

As discussed in chapter 2, many young-earth creationists say that the speed of light was millions of times faster in centuries past, or that continents drifted miles per year; as discussed in chapter 8, some hold to a "canopy" theory that the earth was shrouded in thick clouds until after people had lived on earth for some time. These theories imply that the universe and the world could go through radical changes without any great effect on our lives.

By contrast, the fine-tuning argument of the intelligent design movement says that even the smallest change in a constant of nature like the speed of light would make life in the universe impossible. Similarly, a change in either the composition of the atmosphere, or the rate of energy released in continental drift, would greatly upset the delicate balance of the planet earth, making it impossible for us to live. Not just any old universe, or any old planet, can support life, only one with specially selected parameters: this is evidence for design.

In contrast, the young-earth creationist says that there is nothing particularly special about the way things are; animals could change dramatically, from herbivorous to carnivorous, or the speed of light could become millions of times faster with no major global problems. Thus, they see less design in the universe than the Darwinist.

- *Young-earth creationists engage in scientific practices widely considered unethical by mainstream scientists.*

This sounds like quite an accusation, but I see it as intrinsic to the young-earth science movement. Young-earth creation scientists say that an enormous amount of modern science is wrong, either through a conspiracy or through shared beliefs that lead scientists to unconsciously suppress or alter data. Therefore, young-earth creation scientists must bypass the modern science establishment.

This is not the case with the intelligent design movement. Intelligent design theorists start with the results of modern science and interpret it in a different framework. They do not reject the findings of modern science, just the spin put on it by popularizers of Darwinist evolution. Intelligent design proponents seek to persuade scientists to look at their own data with new eyes.

A few years ago, a major story in the physics world was "cold fusion." Scientists in Utah claimed to have created nuclear reactions in a bottle, using just water and a few special materials, for a tiny fraction of the cost of a nuclear power plant. Clearly, this would have major economic implications. Fortunes were made and lost as venture capitalists dealt in the materials used in the process.

Eventually, the cold fusion claims were disproven to the satisfaction of most scientists (a small group of scientists continue to pursue this effect to this day). In the aftermath, the great majority of scientists felt the original scientists had engaged in unethical, or "pathological," science. What made it pathological was not that they were wrong; many scientists get things wrong all the time. But these scientists bypassed the normal scientific avenues of fact checking and went straight to the public with their claims. When other scientists challenged their claims, the cold fusion advocates went to the law courts to stop them. Most scientists feel that bypassing other scientists to market your scientific claims directly to the public is highly unethical, since the public is not qualified to evaluate scientific claims. Using the law to enforce those claims only adds to the ethical breach.

I tell this story to illustrate that there *is* a "scientific establishment," and the scientific establishment plays a vital role in weeding out nonsense. The scientific establishment gets things wrong sometimes. But we have no choice except to try to persuade that scientific establishment of our views. As F. Scott Fitzgerald is reputed to have said, "No great idea was ever generated by a conference, but a lot of foolish ideas have died there."

Accrediting bodies play an important role in society. We want to avoid quack doctors, homes and food claimed to be safe by fake inspectors, and pastors with no real knowledge of the Bible and theology. Accrediting bodies get it wrong at times: the medical establishment may be too slow to approve new miracle drugs, the Food and Drug administration may approve chemical additives that scare us, and church denominations may lose their grounding in the Bible. But imagine a world (we are not too far from it, perhaps) in which anyone can proclaim herself a doctor, anyone can sell you food claiming it to be safe, and anyone can preach in the name of your church.

The young-earth creation science movement participates in the same type of anarchy of authority in the area of science that the cold fusion scientists did. New findings are often circulated to churches and pronounced on the radio as major finds before scientists from the other side, or even the same side, have had a chance to critique them. Speakers who have no advanced degrees go on the road claiming to be scientists. Christian theologians must weigh carefully the moral implications of having "differing weights and differing measures" (Prov. 20:10 NIV) in the area of science.

I recently came across the path of a man who is a popular creation-science speaker who claims to have a new theory of how the speed of light could change over time. He has no advanced degree in science, and has learned neither general relativity nor quantum electrodynamic field theory, which are the basic starting points of modern cosmology. His argument for the speed of light slowing is very simple. He starts with the electromagnetic energy density equation, which is effectively:

$$U = \frac{E^2}{c^2},$$

In other words, the energy density is proportional to the electric field magnitude squared divided by the speed of light squared. From this he deduces:

$$c^2 = \frac{E^2}{U},$$

Therefore, by his logic, if U increases, c decreases. He believes that as the universe expands, zero-point energy from quantum mechanics is released, which makes the energy density greater.

Any reader with physics or engineering training will find his argument laughable. There is no reason to assume that E stays constant while U increases. By the same argument, the speed of sound will get slower if the sound gets more intense. Readers without any such training will find the above argument mystifying. Perhaps they will even think I am engaging in intellectual snobbery.

To those readers, I submit the following analogy. Suppose a man comes along who speaks no Russian, but who claims to have a new theory of Russian literature based on the view that the Russian word "da" means "dog." Every Russian literature expert you know finds this laughable. Should you take his word above that of the experts because the man is a good Christian man? Should you take his word above that of Christian experts in Russian literature who speak Russian? What would it say about the church's standards if it embraced this type of scholarship?

Can a movement survive that claims to speak truth to the science world, while at the same time it embraces all manner of speakers who contradict basic physics principles? The Christian world is so used to reacting against credentialed scientists that it threatens to fall into gullibility of the highest order. I have heard well-meaning Christian friends tell me seriously all about the benefits of such things as homeopathy or magnet therapy, or about the dangers of electric fields or fluoride, with no regard for the opinions of experts, because they have gotten used to the view that all the experts are members of an establishment that cannot be trusted.

As hard as it may be, we must work to convince the scientific world, not bypass it. This means that we must take the time to learn the basic rules of the secular scientific world, even while we question the unproven assumptions we hear. Many missions experts affirm that to impact a culture, the church must address the top elements of society, lest it be permanently marginalized. Not only is it a good strategy, it also avoids the accusation that the church preys on the weak and ignorant, on those who cannot adequately evaluate the credibility of claims. This was the approach of the apostle Paul, who went to both intellectuals and common people, who was "all things to all people" (1 Cor. 9:22), and who said, "We destroy arguments and every lofty opinion raised against the knowledge of God, and take every thought captive to obey Christ" (2 Cor. 10:5).

Some young-earth creationists do not reject modern science because they take a complete "appearance of age" view in which God created all things a short time ago, but with the perfect appearance of billions of years of history. This view, a type of solipsism, avoids the problems above, but has to deal with the theological problem of God doing things to deceive us.

After reading this discussion, some young-earth creationists may conclude that I am right, that young-earth creationism and intelligent design are incompatible, but it is the intelligent design movement that is intrinsically flawed, and they should distance themselves from it. It may be for the best. As missions experts can attest, having different movements can be a healthy thing. Religious freedom allows us to have different church denominations and parachurch organizations, and while we may feel we could get more done with more unity, where there is disunity, different organizations and movements may actually help us to stop fighting internally and get to the work.

Nonnegotiables

As I discussed in chapter 1, one of the fears many Christians have of old-earth creationism is the "slippery slope": if we allow this long-held belief to change, what might next follow? In response, I will present doctrines that I believe are essential, and are not changed by our position on the age of the earth.

- *Adam was one, real, historical man.*

The parallel of Adam and Jesus in Romans 5 does not allow an interpretation of Adam as merely symbolic:

> But the free gift is not like the trespass. For if many died through one man's trespass, much more have the grace of God and the free gift by the grace of that one man Jesus Christ abounded for many. And the free gift is not like the result of that one man's sin. For the judgment following one trespass brought condemnation, but the free gift following many trespasses brought justification. If, because of one man's trespass, death reigned through that one man, much more will those who receive the abundance of grace and the free gift of righteousness reign in life through the one man Jesus Christ.
>
> Romans 5:15–17

Paul says that through "one man" sin entered into the world, and through "one man," Jesus, we are saved. If we affirm Jesus as one, real historical man, we must take Adam as also one real, historical man. Just as we find ourselves in misery as branches from the root of Adam, so our only hope of salvation is to be engrafted onto the vine of Christ. The genealogy linking all people to Adam also strongly indicates that he was a real man. The authors of the Bible use the genealogies to show that the people in the stories are real.

There is no reason to doubt the miraculous nature of the creation of Adam and Eve. God could have given his spirit to an animal, but the miracles of Genesis 2 seem to be special events that are designed to give us object lessons: first, that humankind is no different from the dust, except for the spirit of God that he gives us, and second, woman and man are intimately connected—woman is "bone of my bone."

- *Noah was one, real, historical man.*

Again, the references to Noah and his genealogy indicate that the story of Noah is meant to be taken historically. The Scriptures also indicate that the flood killed every other human except those on the ark, so that every person today is descended from one of Noah's sons.

This doctrine does not have the same essential role in the Gospel as the role of Adam's sin. The major role of the flood in Scripture is as a type, or symbol, of God's universal final judgment. Every person is not only subject to judgment, without warning, but the forces of nature also continue on in balance, not because of some impersonal law, but only because God holds them back, so he can release them when he chooses.

- *Life in all its diversity was created by sovereign, miraculous acts of God.*

An old-earth view is not synonymous with evolution. Even many millions of years is not enough time to make the design we see occur entirely by random, uncorrelated forces. In particular, although there are certainly various physical attributes that people have in common with animals, the Bible teaches that man and woman were created by a special miracle and did not evolve from primates. This does not mean we are utterly different from primates, but we have a miraculous history.

We do not need to fear similarities with the animals, nor reject the idea that some things have changed over time. But the "image of God" has central importance in the Bible. The fact that we are bearers of the image of God makes humans different from all animals and makes us siblings of Jesus. The miraculous creation from dust is one of God's ways of emphasizing this unique role.

- *The second coming and the final judgment will entail a complete re-creation of the universe and all physical laws.*

Unlike the supposed re-creations that are claimed to have occurred after the fall and the flood, a complete, dramatic change in the physical universe at the new creation is clearly taught in Scripture. Some have argued that the descriptions of Revelation 1–19 represent events that occur in history in this world, but the description of Revelation 21–22 leaves no room for any interpretation except an entirely new creation.

Summing Up

As we have gone through this book, we have touched on very deep themes that run throughout the Bible, such as the very nature of God (Is wrath and "terribleness" essential to his character?), our view of the natural world around us (Is it "twisted?" Should we find joy in it?), and our view of the future (Do we end up where we started, or build something greater by taming the wild things?). The debate over the age of the earth is not just an academic exercise in dating but a very lively debate over the very core themes of the Bible, which relate to our view of all of life.

Is an old-earth view a heretical, "liberal" viewpoint? A human-centered compromise? Should people who adhere to these theological implications be drummed out of the orthodox church, or forbidden to preach? I hope that those who are not convinced by my arguments will at least find them to be *biblical* arguments, grounded in study of the entire Word of God.

God's Word and God's universe may contain many surprises for us.

> Therefore every teacher of the law who has been instructed about the kingdom of heaven is like the owner of a house who brings out of his storeroom new treasures as well as old.
>
> Matthew 13:52 NIV

Review Questions

1. Have you forced a view of God as "nice" onto the Bible?
2. Do you feel that it's not "grand" enough that Adam and Eve simply got kicked out of the Garden, without any great change in the rest of the world, or that the flood of Noah was only tens of feet deep? Is that a valid argument against any Bible interpretation?
3. Should the church demand that people claiming to speak as experts about science take the time to get credentials in science fields, even if they disagree with much of what they are learning? By comparison, is it "snobbery" to want the electrician who works on your house to be certified? Does bypassing the "establishment" violate the ethical principle of not having varying weights and measures?

Appendix

A Literal Translation of Genesis

Much of our debate about these issues is influenced by the English translations we read. For example, reading the word *earth* brings to mind our modern images of a globe as seen from outer space. The ancient Hebrew reader would have had no such image—the word translated as *earth* is generic, much closer to our word *land* in all its ambiguity. Other words, like the word sometimes translated *expanse* or *firmament*, have no exact equivalent in English, and these translations bring to mind nothing concrete at all. On the other hand, words like *covenant* and *consecrate* had very specific meanings in Hebrew, which may be lost on modern readers, but no better word exists in English.

In the following translation, I have tried to present as much as possible the original sense of the text. As much as possible, the original sense of the words has been used, such as *flyers* or *creepers* instead of *birds* or *animals*. This is valuable because ancient categories of living things did not correspond to our modern ones; for example, flying bugs belonged with birds in the category of "flyers" and frogs, rabbits, and bugs all belonged in the category of "creepers." Using words close to the original meaning also helps us to see some of the poetic alliteration in the original text, such as "creepers that creep." I have also maintained the principle as much as possible of using one English word or phrase for one Hebrew word, so that it is clear when the Hebrew words are the same in the text and when they are not.

There are other more subtle biases in modern translations, such as the use of many different verb tenses, while in Hebrew there were fewer ways to indicate the relative time of events. As much as possible, I have adhered to the principle of one English verb tense for one Hebrew verb tense. This limits the translation to two past tenses: simple past ("perfect"), e.g., "he ran" and connected past tense ("waw-consecutive"), translated here by adding "and so" in every case, e.g., "and so he ran." This repeated form can sound repetitious to modern readers, and therefore most modern translations substitute other words such as "then" or "next," but in so doing they force a particular view of the time sequence. The generic form "and so" has the meaning of adding another thing, without a definite cause and effect or time sequence implied; it is the standard phrase of story telling, when a new element is to be added. Occasionally time-sequence words such as *after* or *until* do appear in the Hebrew, and in such cases I include them.

Quotation marks also are a fairly recent invention and do not appear in the original text. I have also not added titles, but instead have used the "titles" that appear in the text, which clearly have the sense of "the following is the account of the generations of . . ." I have also included the * character, indicating a dramatic break, where it occurs in the text, and ** where a full-stop character, indicating a longer dramatic break, occurs in the original text. For the sake of not wearying the English reader, I have broken the text into paragraphs between these marks, but these paragraphs are not indicated by the text. I have also tried to include the definite particle "the" only when it appears in the text, and in every case when it does.

The text is necessarily awkward in places, as any literal translation must be. Yet it also has a certain majesty, or gravitas, that is lost in modern translations. One feels, in this translation, the ancientness of the text.

I am not a professional Hebrew scholar, and so I must acknowledge several aids, including the *Interlinear Bible* and *A Literal Translation of the Bible*, edited by Jay P. Green, Sr. (Peabody, Massachusetts: Hendrickson Publishers); the *Biblia Hebraica Stuttgartensia*, edited by Karl Elliger, William Rudolph, et al., 4th corrected edition edited by Adrian Schenker (Stuttgart, Germany: Deutsche Bibelgesellschaft); the *Groves-Wheeler Westminster Hebrew Morphology* (Philadelphia: Westminster Theological Seminary); Strong's Hebrew and Chaldee dictionary (McLean, Virginia: MacDonald Publishing); and input from several Bible scholars including Jack Collins of Covenant Seminary, St. Louis, and Yaakov Ulano of Israel. Electronic Bible software such as Accordance from OakTree Software has made word study far easier than it once was.

Many commentators have discussed how these chapters represent the preamble to a covenant document, declaring the relation of the participants. I end this translation in Genesis 12:7, because there begins the story of the covenant of Abraham, the founding story of redemptive history of the nation of Israel.

The First Book of Moses

¹:¹In the beginning God created the heavens and the land. ²And the land was formless and empty, darkness was on the face of the deep, and the Wind of God was brooding over the waters. ³And so God said, Let there be light. And so there was light. ⁴And so God saw that the light was good, and so he separated the light from the darkness. ⁵And so God named the light day, and the darkness he named night. And so there was evening, and there was morning, a first day.

**

⁶And so God said, Let there be a shiny surface between the waters to separate waters from waters. ⁷And so God made the shiny surface and separated the waters under the shiny surface from the waters above it, and so it was thus. ⁸And so God named the shiny surface the heavens, and so there was evening, and there was morning, a second day.

**

⁹And so God said, Let the waters under the heavens be gathered to

one place, and dry ground appear, and so it was thus. ¹⁰And so God named the dry ground land, and the gathered waters he named seas, and so God saw that it was good. ¹¹And so God said, Let the land sprout seed-bearing plants and trees on the land that bear fruit with seed in it, according to their various kinds, and so it was thus. ¹²And so out of the land went plants bearing seed according to their kinds and trees bearing fruit with seed according to their kinds, and so God saw that it was good. ¹³And so there was evening, and there was morning, a third day.

**

¹⁴And so God said, Let there be lights in the shiny surface of the heavens to separate the day from the night, and they will serve as signs to mark appointed times and days and years. ¹⁵And they will be lights in the shiny surface of the heavens to give light on the land. And so it was thus. ¹⁶And so God made two great lights, the greater light to govern the day and the lesser light to govern the night, and the stars. ¹⁷And so God set them in the shiny surface of the heavens to give light on the land; ¹⁸and to govern the day and the night, and to separate light from darkness, and so God saw that it was good. ¹⁹And so there was evening, and there was morning, a fourth day.

**

²⁰And so God said, Let the waters swarm with swarmers, living souls, and let flyers fly above the land across the face of the shiny surface of the heavens. ²¹And so God created the great reptile-monsters and all the living souls, which creep, which swarmed the waters, according to their kinds, and every winged flyer according to its kind, and so God saw that it was good. ²²And so God blessed them, saying, You will be fruitful and increase in number and fill the waters in the seas, and the flyers will increase on the land. ²³And so there was evening, and there was morning, a fifth day.

**

²⁴And so God said, Let living souls go out from the land according to their kinds, beasts, creepers and living things of the land, each according to its kind. And so it was thus. ²⁵And so God made the living things of the land according to their kinds, the beasts according to their kinds, and all the creepers on the ground, according to their kinds, and so God saw that it was good. ²⁶And so God said, Let us make a human[1] in our image, in our likeness, and let them rule over the swimmers of the sea and the flyers of the heavens, over the beasts, over all the land, and over all the creepers that creep on the land. ²⁷And so God created the human in his image, in the image of God he created him, male and female he created them. ²⁸And so God blessed them, and so God

said to them, You will be fruitful and increase in number, and you will fill the land and subdue it, and you will rule over the swimmers of the sea and the flyers of the heavens and over every living thing that creeps on the land. [29]And so God said, Behold, I gave you every plant for sowing seed on the face of the whole land and every tree that has fruit with seed in it; it will be yours for food; [30]and to all the living things of the land and all the flyers of the heavens and all the creepers on the land, everything with a living soul, every green plant for food, and so it was thus. [31]And so God saw all that he made, and behold, it was very good, and so there was evening, and there was morning, the sixth day.

**

[2:1]And so the heavens and the land were completed, and all the host of them. [2]And so with the seventh day God completed the work he did, and so in the seventh day he ceased from all his work. [3]And so God blessed the seventh day and consecrated it, because on it God ceased from all his work of doing what he created.

**

[4]*These are the generations of the heavens and the land, with the creating of them, in the day of the I AM God making the land and the heavens.*

[5]And before every shrub of the field was on the land and before every plant of the field would spring up, because the I AM God sent no rain on the land, and there was no human to work the humus; [6]and mist would go up from the land and would give drink to the whole face of the ground; [7]and so the I AM God formed the human from the dust of the ground and blew into his nose the breath of life, and so the human was a living soul. [8]And so the I AM God planted a garden in Eden, in the east, and so he put the human he formed there. [9]And so the I AM God made all kinds of trees grow out of the ground, trees to please the eye and good for food, and the tree of life was in the middle of the garden, and the tree of the knowledge of good and evil; [10]and a river going out from Eden to give drink to the garden, and from there it would separate to be four heads: [11]the name of the first, Pishon, circling all the land of Havilah, with gold there; [12]and the gold of that land is good, and there is aromatic resin and onyx; [13]and the name of the second river, Gihon, circling around the whole land of Cush; [14]and the name of the third river, Hiddekel, going to the east of Assyria; and the fourth river, Euphrates.

[15]And so the I AM God took the human to rest him in the Garden of Eden, to work it and protect it. [16]And so the I AM God commanded the human, saying, You may eat of all the trees for eating

in the garden, [17]and you may not eat from the tree of the knowledge of good and evil, because in the day you eat of it you will die the death.

[18]And so the I AM God said, It not being good for the human to be alone, I will make a helper fit for him. [19]And so the I AM God formed from the humus all the living things of the field and all the flyers of the heavens, and so he brought them to the human, to see what he would name them, and whatever the human called each living soul, that was its name. [20]And so the human called all the beasts by name, the flyers of the heavens and all the living things of the field, and for a human no helper fit for him was found. [21]And so the I AM God caused a deep sleep to fall on the human, and so he slept, and so he took one rib from him and closed up the place with flesh. [22]And so the I AM God built the rib which he took out of the human into a woman, and so she came to the human. [23]And so the human said, At last, bone of my bones and flesh of my flesh. She shall be called woman, for she was taken out of man. [24]For this reason a man will leave his father and mother and attach to his woman, and they will become one flesh. [25]And so the two of them were naked, the human and his woman, and they would not feel shame.

[3:1]And the snake was more crafty than any living thing of the field which the I AM God made, and so he said to the woman, Indeed, God said, you are not to eat from every tree in the garden? [2]And so the woman said to the snake, We are to eat from tree fruit in the garden, [3]and from the fruit of the tree which is in the middle of the garden, God said, you are not to eat from it, and you are not to touch it, lest you die. [4]And so the snake said to the woman, You will not die the death; [5]because God knows that when you eat of it, your eyes will be opened, and you will be like God, knowing good and evil. [6]And so the woman saw that the fruit of the tree was good for food and pleasing to the eye, and also desirable for gaining wisdom, and so she took from the fruit and so she ate it, and so she also gave some to her man, who was with her, and so he ate it. [7]And so the eyes of both of them were opened, and they knew they were naked, and so they sewed fig leaves together and made girdles for themselves.

[8]And so they heard the sound of the I AM God walking in the garden in the wind of the day, and so the human and his woman hid from the I AM God among the trees of the garden. [9]And so the I AM God called to the human, and so he said to him, Where are you? [10]And so he said, I heard you in the garden, and so I feared because I was naked, and so I hid. [11]And so he said, Who told you

that you were naked? Did you eat from the tree that I commanded you, except eating it, to eat? [12]And so the human said, The woman you put here with me, she gave me from the tree to eat, and so I ate. [13]And so the I AM God said to the woman, What is this you did? and the woman said, The snake deceived me, and I ate. [14]And so the I AM God said to the snake, Because you did this, cursed are you above all the beasts and all the living things of the field; you will crawl on your belly and you will eat dust all the days of your life. [15]I will set enmity between you and the woman, and between your seed and her seed; he will crush your head, and you will strike his heel.

**

[16]To the woman he said, Great will be your pains in pregnancy, with pain you will give birth to sons; your desire will be to control your man, and he will rule over you.

**

[17]To Human he said, Because you heard the voice of your woman, and so you ate from the tree about which I commanded you, saying you are not to eat of it, cursed is the humus because of you; through painful toil you will eat its produce all the days of your life. [18]Thorns and thistles will sprout for you, and you will eat the plants of the field. [19]With the sweat of your face you will eat your food until you return to the humus, because from it you were taken, because you are dust and to dust you will return. [20]And so the human called the name of his woman Life-giver, because she was the mother of all the living. [21]And so the I AM God made robes of skin for human and his woman and clothed them.

**

[22]And so the I AM God said, Behold, the human was like one of us, knowing good and evil, and now lest he send out his hand and take also from the tree of life and eat, and live forever: [23]and so the I AM God sent him out from the garden of Eden to work the humus from which he was taken. [24]And so he drove the human out, and so angels dwelled on the east side of the Garden of Eden, and a flaming sword to turn to protect the way to the tree of life.

**

[4:1]And the human knew his woman Life-giver, and so she became pregnant and gave birth to Spear, and so she said, I obtained ownership[2] of a man from the I AM; [2]and so she repeated, giving birth to his brother Vapor; and so Vapor was shepherding flocks, and Spear working the humus. [3]And so it was at the end of days, and so Spear came with the fruit of the humus as an offering to the I AM. [4]And Vapor also came, with fat portions from the firstborn of his flock, and so the I AM gazed on Vapor and his offering. [5]And on Spear and his offering he did not gaze, and Spear blazed greatly, and so his face fell. [6]And so the I AM said to Spear, Why did you burn, and why did your face fall? [7]If you do what is right, will you not be accepted? And if you do not do what is right, sin crouches at your door; its desire will be to control you, and you must rule over it.

[8]And so Spear said to his brother Vapor, and so it was that they were in the field, and so Spear rose up against his brother Vapor, and so he killed him. [9]And so the I AM said to Spear, Where is your brother Vapor? And so he said, I don't know; am I my brother's protector? [10]And he said, What did you do? The voice of your brother's blood is crying out to me from the humus. [11]And now, cursed are you on the humus, which opened its mouth to receive your brother's blood from your hand; [12]because when you work the humus, it will no longer yield its strength for you, and you will be wandering and mourning to and fro in the land.

[13]And so Spear said to the I AM, Punishment more than I can carry! [14]Behold, this day you drove me out over the face of the humus, and I will be hidden from your face; and I will be wandering and mourning to and fro in the land, and anyone who finds me will kill me. [15]And so the I AM said to him, Not so, if anyone kills Spear, he will be avenged seven times; and so the I AM put a mark on Spear, to except him from killing by all who find him. [16]And so Spear went out from the presence of the I AM, and so dwelled in the land of Nod, east of Eden. [17]And so Spear knew his woman, and so she became pregnant, and so gave birth to Initiated; and so he was building a city, and so called the name of the city after the name of his son, Initiated . . .

[4:25]And so Human knew his woman again, and so she gave birth to a son, and so called his name Substitute: for God has appointed to me another seed instead of Vapor, whom Spear killed. [26]And to Substitute also was begotten a son, and so he called his name Woeful; then men began to call on the name of the I AM.

**

[5:1]***This is the book of the generations of Human, in the day of God creating Human, in the likeness of God he made him. [2]Male and female he created them, and so he blessed them, and so he called his name Human in the day of creating.***

**

[3]And so Human lived a thirty-and-hundred-year, and so he begot a son in his likeness, in his image, and so he called him Substitute. [4]And so the days of Human after begetting Substitute were an eight-hundred-year, and so he begot sons and daughters. [5]And so all the days

of Human which he lived were a nine-hundred-year and a thirty-year; and so he died . . .

**

$^{5:28}$And so Lamech lived a two-and-eighty-year and a hundred-year, and so he begot a son. ^{29}And he called his name Consolation, saying, He will console us from labor and from painful toil of our hands from the ground the I AM cursed. ^{30}And so the days of Lamech after begetting Consolation were a five-and-ninety-year and a five-hundred-year, and so he begot sons and daughters. ^{31}And so all the days of Lamech were a seventy-seven-year and a seven-hundred-year; and so he died.

**

$^{5:32}$And so Consolation was a five-hundred-year son, and so he begot Honor, Heat, and Growth. $^{6:1}$And so the human began to multiply on the land, and daughters were begotten to them. ^{2}And so the sons of God saw the daughters of the human, because they were beautiful, and so they took women from all that they chose. ^{3}And so the I AM said, My Wind will not contend ages with a human, who is merely flesh, and his days will be a hundred-and-twenty-year. ^{4}And the giants were in the land in those days, and also afterward, because the sons of God would come to the daughters of the human and they begot by them; they were the mighty of ages, men of the name.

**

^{5}And so the I AM saw that evil of the human was great in the land, and every inclination of his heart's thoughts only evil all the day. ^{6}And so the I AM consoled himself because he made the human on the land, and so it pained him to his heart. ^{7}And so the I AM said, I will wipe the human, whom I created, from the face of the humus, from human to beast to creeper to flyer of the heavens, because I console myself that I made them. ^{8}And Consolation found grace in the eyes of the I AM.

**

^{9}***These are the generations of Consolation, a sound man, complete among his generation, who walked with God.*** ^{10}***And so Consolation begot three sons, Honor, Heat and Growth.***

^{11}And so the land was ruined before the face of God, and so the land was filled with violence. ^{12}And so God saw the land, and behold, it was ruined, because all flesh on the land ruined its way.

**

^{13}And so God said to Consolation, An end to all flesh comes before me, because the land was filled with violence from them; and behold, I will ruin them with the land. ^{14}Make a box of cypress wood; you will make dwellings in it and cover it with pitch inside and out. ^{15}And this is how you will make it: three-hundred-cubit length of the box, fifty-cubit width, thirty-cubit height. ¹⁶You will make a roof for the box and finish the box to one cubit above, and you will place a door in the side of the box and you will make lower, second and third decks. ¹⁷And behold, I will bring floodwaters on the land to ruin all flesh with the wind of life under the heavens; all on the land will die. ¹⁸And I will establish my covenant with you, and you will come to the box, you and your sons and your woman and your sons' women with you. ¹⁹And from all the living things of all flesh, two from all are to come to live in the box with you; they will be male and female. ²⁰From every kind of flyer, from every kind of beast and from every kind of creeper of the ground will come to you to live. ²¹You will take every kind of food which is to be eaten and gather it as food for you and for them. ²²And so Consolation did everything that God commanded him.

**

⁷:¹And so the I AM said to Consolation, Go into the box, you and your whole house, because I saw you as sound in this generation. ²You will take seven by seven of every clean beast, a man and its woman, and two of every unclean beast, a man and his woman, ³and also seven by seven of every flyer, male and female, to keep their seed alive on the face of the land. ⁴Because after seven days I send rain on the land a forty-day and a forty-night, and I will wipe from the face of the humus every growing thing which I made. ⁵And so Consolation did all that the I AM commanded him.

⁶Consolation was a six-hundred-year son, and the floodwaters were on the land. ⁷And so Consolation and his sons and his woman and his sons' women came into the box to escape the waters of the flood. ⁸From clean beasts and unclean beasts, and from flyers and from all creepers on the humus, ⁹two by two, male and female, they came into the box with Consolation, as God commanded Consolation. ¹⁰And so after the seven days the floodwaters were on the land. ¹¹In the six hundredth year of Consolation's life, on the seventeenth day of the second month, on this day, all the springs of the great deep split open and the gates of the heavens were opened. ¹²And rain fell on the land a forty-day and a forty-night. ¹³On that same day Consolation and his sons, Honor, Heat and Growth, together with his woman and the women of his three sons, came to the box, ¹⁴they and every living thing according to its kind, every beast according to its kind, every creeper creeping on the land according to its kind and every flyer according to its kind, everything with wings. ¹⁵And so they came into the box with Consolation, two by two of all flesh that have the wind

of life. ¹⁶And coming were male and female of all flesh, as God commanded Consolation, and so the I AM shut them in. ¹⁷And so the flood was a forty-day on the land, and so the waters multiplied and lifted the box and raised it high above the land. ¹⁸And so the waters prevailed and multiplied greatly on the land, and so the box moved on the face of the waters. ¹⁹And the waters prevailed greatly on the land, and so covered all the high hills under the whole heavens. ²⁰The waters prevailed a fifteen-cubit upward and so covered the hills. ²¹And so all flesh that creeps on the land died, with the flyer, with the beast, with the living thing, with all the swarmers that swarm on the land, and all of the human. ²²Everything on dry land that had the wind of life in its nose died. ²³And so every growing thing on the face of the humus was wiped out, from human to beast to creeper to flyer of the heavens; and they were wiped from the land, and only Consolation was left, and those with him in the box. ²⁴And so the waters prevailed on the land for a hundred and fifty days.

⁸:¹And so God remembered Consolation and all the living things and the beasts that were with him in the box, and so God passed by a wind over the land, and the waters lowered. ²And so the springs of the deep and the gates of the heavens were closed, and so the rain from the heavens was stopped. ³And so the waters returned, walking, returning, and so the water decreased on the land at the end of a fifty-and-hundred-day. ⁴And so the box rested in the seventh month, on day seventeen of the month, on the hills of Ararat. ⁵And the waters were walking, decreasing, until the tenth month; with the tenth, on one of the month, the heads of the hills were seen. ⁶And so it was after a forty-day, and so Consolation opened the window he made in the box. ⁷And so he sent out a raven, and so it went out and returned until the drying of the water from the land. ⁸And so he sent out a dove with him to see the water reducing from the face of the ground. ⁹And the dove found no rest for its hand-foot, and so returned to the ark, because of the water over all the face of the land; and so he sent out his hand and took the dove and so it came back to him in the box. ¹⁰And so he waited another seven days, and so he repeated sending out the dove from the box. ¹¹And so the dove came to him in the evening time, and behold, a plucked olive leaf in its mouth, and so Consolation knew that the water reduced from on the land. ¹²And so he waited again another seven days, and so sent the dove out, and it did not repeat returning to him again. ¹³And so it was, with the one-and-six-hundred year, month one, the

first, the water dried from on the land; and so Consolation turned aside the covering from the box and so saw, and behold, the face of the humus dried. [14]And with the second month, with day twenty-seven, the land dried up.

**

[15]And God spoke to Consolation, saying, [16]Go out of the box, you and your woman and your sons and your sons' women, [17] every kind of living thing that is with you from all flesh, of the flyers, of the beasts, of all the creepers creeping on the land; they will go out with you to swarm on the land, to be fruitful, and to multiply on the land. [18]And so Consolation went out, together with his sons and his women and his sons' women. [19]All the living things, all the creepers, and all the flyers, everything that creeps on the land, in families, went out of the box.

[20]And so Consolation built an altar to the I AM, and so took from all the clean beasts and the clean flyers, and so sent up burnt offerings on the altar. [21]And so the I AM smelled the pleasing smell, and so the I AM said to his heart, I will not reduce the humus again because of the human because every inclination of the human's heart is evil from youth, not repeating again killing everything living, as I did. [22]All the days of the land, seedtime and harvest, cold and heat, summer and winter, day and night will not cease again.

[9:1]And so God blessed Consolation and his sons, saying to them, Be fruitful and multiply and fill the land. [2]And fear of you and dread of you will be in all the living things of the land, in all the flyers of the heavens, in all which creep on the humus, and in all the swimmers of the sea, all given into your hand. [3]Every living creeper will be food for you; like the green plants, all is given to you. [4]But the flesh with its soul, its blood, in it, you are not to eat. [5]But your blood, your souls, I will seek; from the hand of every living thing I will be seek it, and from the hand of the human, from the hand of the brother of a man, I will seek the soul of the human. [6]Pouring out blood of the human, by a human his blood will be poured out, because the human was made in the image of God. [7]And you, be fruitful and multiply, and swarm on the land and multiply in it.

**

[8]And so God said to Consolation and to his sons with him, [9]And behold, I establish my covenant with you and with your seed after you, [10]and with every living soul that was with you, with flyer, beast, and every living thing of the land with you, with all going out of the box with you, every living thing in the land. [11]I will establish my covenant with you, and all flesh will not be cut off

by the waters of a flood again, and a flood will not ruin the land again. [12]And so God said, This is the sign of the covenant that I set between me and you and every living soul with you, to generations of the ages. [13]I set my bow in a cloud, and it will be a sign of the covenant between me and the land. [14]And my clouding of a cloud will be over the land, and the bow in the cloud will be seen, [15]And I will remember my covenant between me and you and all living souls of all flesh, and the waters will not be a flood to ruin all flesh. [16]And the bow will be in the cloud, and I will see it, remembering the covenant of the ages between God and all living souls of all flesh on the land. [17]And so God said to Consolation, this is the sign of the covenant which I established between me and all flesh on the land.

**

[18]And so the sons of Consolation who went out of the box were Honor, Heat and Growth, and Heat was the father of Servant. [19]Three these sons of Consolation, and from these scattered all of the land.

[20]And Consolation began to be a man of the humus, and so he planted a vineyard. [21]And so he drank some of its wine, and so he became drunk, and he was exposed inside his tent. [22]And so Heat, the father of Servant, saw his father's nakedness and declared it to his two brothers outside. [23]And so Honor and Growth took the garment and laid it across both of their shoulders, and so they walked backward and covered their father's nakedness, and their faces were backward, and they did not see their father's nakedness.

[24]And so Consolation awoke from his wine, and so he knew what his youngest son did to him. [25]And so he said, Cursed is Servant, a slave of slaves he will be to his brothers. [26]He also said, Blessed is the I AM, the God of Honor; Servant will be the slave of Honor. [27]Let God enlarge the territory of Growth, and let Growth live in the tents of Honor, and let Servant be his slave.

[28]And so Consolation lived after the flood a three-hundred year and a fifty-year. [29]And all the days of Consolation were a nine-hundred-year and a fifty-year; and so he died.

**

[10:1]*These are the generations of Honor, Heat and Growth, Consolation's sons. And so they begot sons after the flood.*

[21]. . . And Honor also begot sons, he who was the father of all the sons of Eber, whose older brother was Growth. [22]The sons of Honor: Elam, Asshur, Arphaxad, Lud and Aram.[3] [23]And the sons of Aram: Uz, Hul, Gether and Meshech. [24]And Arphaxad begot Shelah, and Shelah begot Eber. [25]And to Eber were begotten two sons; the

name of one was Split, because in his days the land was split; and the name of his brother was Little.

[31]. . . These are the sons of Honor, by their families, by their tongues, in their lands, by their nations. [32]These the families of Consolation's sons, in their generations, in their nations; and from these the nations separated in the land after the flood.

**

[11:1]And so there was in all the land one lip and one speech. [2]And so it was, journeying in the east, and so they found a plain in the land of Shinar, and so they settled there. [3]And so a man said to his companion, Come, let us make bricks and let us bake them baked; and so was to them brick for stone, and tar for mortar. [4]And so they said, Come, let us build ourselves a city and a tower and its head in the heavens, and let us make a name for ourselves, lest we be scattered over the face of the whole land.

[5]And so the I AM went down to see the city and the tower that the sons of the human built. [6]And so the I AM said, Behold, they are one people, and they have one tongue, and this is the beginning of what they will do, and nothing they purpose to do will be impenetrable for them. [7]Come, let us go down and mix their lip so a man will not hear the lip of his companion. [8]And so the I AM scattered them from there over all the land, and so they stopped building the city. [9]Therefore its name was called Mixed-up, because there the I AM mixed the lip of the whole land, and from there the I AM scattered them over the face of the whole land.

**

[10]*These are the generations of Honor.*

Honor was a hundred-year son, and so he begot Arphaxad, two years after the flood. [11]And so Honor lived after he begot Arphaxad a five-hundred-year, and so he begot sons and daughters.

**

[25]. . . And so Nahor lived after begetting Terah[4] a nineteen-year and a hundred-year, and so he begot sons and daughters.

*

[26]And so Terah lived a seventy-year, and so he begot High Father, Sleeper, and Mountaineer.

[27]*These are the generations of Terah.*

Terah begot High Father, Sleeper and Mountaineer; and Mountaineer begot Covered. [28]And so Mountaineer died before the face of his father Terah, in Ur of the Chaldeans, in the land of his birth. [29]And so High Father and Sleeper both took women. The name of High Father's woman was Princess, and the name of Sleeper's woman was Queen, the daughter of Mountaineer the father of Queen and the father of Watcher. ³⁰And so Princess was barren and she had no child. ³¹Terah took his son High Father, his grandson Covered, son of Mountaineer, and his daughter-in-law Princess, the woman of his son High Father, and so they set out from Ur of the Chaldeans, walking to the land of Servant; and so they came to Mountaineer, and so they settled there. ³²And so the days of Terah were a five-year and a two-hundred-year, and so Terah died with Mountaineer.

**

¹²:¹. . . And so the I AM said to High Father, Walk from your land, from your homeland, and from your father's house, to the land I will make seen. ²And I will make you into a great nation and I will bless you and I will make your name great, and you will be a blessing. ³I will bless those blessing you, and those reducing you I will curse, and in you all the families of the humus will be blessed.

⁴And so High Father walked, as the I AM spoke to him, and Covered walked with him; and High Father was a son of five years and a seventy-year in going out from Mountaineer. ⁵And he took his woman Princess, and Covered, son of his brother, and all the possessions they accumulated and the souls they made in Mountaineer, and so they went out to walk to the land of Servant, and so they came to the land of Servant. ⁶And so High Father traveled through the land as far as the place of the great tree of Moreh at Shechem. And the people of Servant were then in the land.

⁷And so the I AM made himself seen to High Father, and so he said, To your seed I will give this, the land. And so he built an altar to the I AM, who was seen to him.

Notes

Chapter 1 Starting Assumptions

1. Adrian L. Melott, "Intelligent Design Is Creationism in a Cheap Tuxedo," *Physics Today* 55 (June 2002): 48.

2. Quotes by Luther and Wesley are from Andrew D. White, *A History of the Warfare of Science with Theology in Christendom* Vol. 1 (Appleton, NY: Prometheus Books, 1932), 126, 128. J. Dillenberger, in *Protestant Thought and Natural Science* (New York: Doubleday, 1960), argues that the quote by Luther may be spurious but Wesley clearly did teach against Copernicus. Calvin clearly taught the Ptolemaic system in several places in his *Commentaries*, e.g. in his commentary on Psalm 93:1; he also insisted that the Scriptures required that the Moon must give off light and not only reflect it, since it is called a "lesser light" (*Commentary on Genesis*, v. 1:15).

3. Owen Gingerich, "The Galileo Affair," *Scientific American* 247 (August, 1982): 132. Reprinted in Owen Gingerich, *The Great Copernicus Chase and Other Adventures in Astronomical History* (Cambridge: Cambridge University Press, 1992), 105.

4. Gleason Archer, *Encyclopedia of Bible Difficulties* (Grand Rapids: Zondervan, 1982), 286.

5. See, e.g., P. Harrison, *The Bible, Protestantism, and the Rise of Natural Science* (Cambridge: Cambridge University Press, 1998).

Chapter 2 The Scientific Case

1. E. Harrison, "Newton and the Infinite Universe," *Physics Today* 39 (February 1986): 24. For a basic discussion see David W. Snoke, *Natural Philosophy: Physics and Western Thought* (Colorado Springs: Access Research Network, 2003), section 8.6.

2. For a discussion of this effect, see David W. Snoke, *Natural Philosophy: A Survey of Physics and Western Thought*, section 8.12.

3. P. Dirac, in *The Physicist's Conception of Nature*, ed. Jagdish Mehra (Boston: Dordrecht, 1973), 45.

4. For a discussion of large-numbers coincidences, see P. C. W. Davies, *The Accidental Universe* (Cambridge: Cambridge University Press, 1982); J. Barrow and F. Tipler, *The Anthropic Cosmological Principle* (Oxford: Oxford University Press, 1988); and Hugh Ross, "Big Bang Model Refined by Fire," in W.A. Dembski, ed., *Mere Creation* (Downer's Grove, IL: InterVarsity, 1998).

5. For example, see Thomas Barnes, "Evidence Points to a Recent Creation," *Christianity Today* (October 8, 1982), 34–36; Henry Morris and Gary Parker, *What is Creation Science?* (San Diego: Creation-Life Publishers, 1982), 247–57; W.T. Brown, Jr., "The Scientific Case for Creation: 116 Categories of Evidence," *Bible-Science Newsletter* (June-July-August 1984). In 1993, A. Snelling, and D. Rush, on the basis of reexamination of modern measurements, concluded that young-earth creationists should no longer use moon dust as an argument: "Moon Dust and the Age of the Solar System," *Creation Ex Nihilo Technical Journal* 7, no. 1 (1993): 2–42. Many young-earth creationists still appeal to this argument, however. Various "blogs" have compiled refutations of young-earth evidence, e.g. http://www.infidels.org/library/modern/dave_matson.

6. See, e.g., P.G. Phillips, "The 15.7 Light-year Universe," *Perspectives on Science and Christian Faith* 40 (March 1988): 21; cf. John Whitcomb and Henry Morris, *The Genesis Flood* (Philadelphia: Presbyterian and Reformed, 1961), 370; G. Mulfinger, Jr., "Reviews of Creationist Astronomy," *Creation Science Research Quarterly* 10 (1973): 174; R. Niesson, "Starlight and the Age of the Universe," *Impact* 121 (July 1983); and A.J. Monty White, *How Old is the Earth?* (Welwyn, England: Evangelical Press, 1985).

7. I. Alexander et al., *Geology* 29 (2001): 483.

8. Don Stoner, *A New Look at an Old Earth* (Eugene, OR: Harvest House, 1997), 84–87.

9. For a general review of geological data, see D. Wonderly, *Neglect of Geologic Data: Sedimentary Strata Compared with Young-Earth Creationist Writings* (Hatfield, PA: Interdisciplinary Biblical Research Institute, 1987).

10. I have not addressed the subject of radiometric dating, a common target of young-earth creationists. For a review of the present status of radiometric dating, see Roger Wiens, "Radiometric Dating: A Christian Perspective," available at http://www.asa3.org/ASA/resources/Wiens.html.

11. Robert Shapiro, *Origins: A Skeptic's Guide to the Creation of Life on Earth* (New York: Summit Books, 1986).

12. Michael J. Behe, *Darwin's Black Box: The Biochemical Challenge to Evolution* (New York: Free Press, 1996).

13. Phillip E. Johnson, *Darwin on Trial* (Downer's Grove, IL: InterVarsity, 1993).

14. For a general review see W. Dembski, ed., *Mere Creation* (Downer's Grove, IL: InterVarsity, 1998).

Chapter 3 The Biblical Case I: Animal Death

1. St. Augustine, *The Literal Meaning of Genesis*, vol. 1, J. H. Hammand, trans. (New York: Newman Press, 1982), 92.

2. Martin Luther, *Lectures on Romans*, W. Pauck, trans. (Philadelphia: Westminster Press, 1961), 238.

3. St. Augustine, *Literal Meaning of Genesis*, 94.

Chapter 4 The Biblical Case II: The Balance Theme in Scripture

1. The translation as "whale" (KJV) is nowhere supported by Scripture. The creature that swallows Jonah is a "fish" ("dag" in Hebrew).

2. Cornelius G. Hunter, *Darwin's God: Evolution and the Problem of Evil* (Grand Rapids: Brazos Press, 2001). This book reviews at length the issue of natural evil as treated by Christians in the past two centuries, including Christian evolutionists.

Chapter 5 The Biblical Case III: The Sabbath

1. See, e.g., Meredith Kline, "Space and Time in the Genesis Cosmogony," *Perspectives on Science and Christian Faith* 48 (March 1996): 2–15.

2. R.C. Newman and H.J. Eckelmann, *Genesis One and the Origin of the Earth* (Downer's Grove, IL: InterVarsity, 1977).

3. Hugh Ross, *The Fingerprint of God* (New Kensington, PA: Whitaker House, 1989); *Creation and Time: A Biblical and Scientific Perspective on the Creation-Date Controversy* (Colorado Springs: NavPress, 1994); *A Matter of Days* (Colorado Springs: NavPress, 2004).

Chapter 6 Concordantist Science

1. Newman and Eckelmann, *Genesis One and the Origin of the Earth*.

2. Hugh Ross, *The Fingerprint of God*; *Creation and Time*; and *A Matter of Days*.

3. John Wiester, *The Genesis Connection* (Hatfield, PA: Interdisciplinary Biblical Research Institute, 1992).

4. Gerald Schroeder, *The Science of God* (New York: The Free Press, 1997).

5. Guillermo Gonzalez and Jay W. Richards, *The Privileged Planet: How Our Place in the Cosmos is Designed for Discovery* (Washington, D.C.: Regnery Publishing, 2004), 65–79.

6. David W. Snoke, "Toward a Unified View of Science and Theology," *Perspectives on Science and Christian Faith* 43 (September 1991): 166–73; "The Problem of the Absolute in Evidential Epistemology," *Perspectives on Science and Christian Faith* 47 (March 1995): 3; "The Apologetic Argument," *Perspectives on Science and Christian Faith* 50 (June 1998): 108–122; "In Favor of God-of-the-Gaps Reasoning," *Perspectives on Science and Christian Faith* 53 (September 2001): 152–158.

7. "Toward a Unified View of Science and Theology."

8. P.L. Maier, *A Skeleton in God's Closet* (Nashville: Thomas Nelson, 1994).

9. H.J. Van Till, "Is the Creation a 'Right Stuff' Universe?" *Perspectives on Science and Christian Faith* 54 (December 2002): 232–39.

10. Paul Davies, *The Accidental Universe* (Cambridge: Cambridge University Press, 1982). Also J.D. Barrow and F.J. Tipler, *The Anthropic Cosmological Principle* (Oxford: Oxford University Press, 1988), chapters 5–6.

11. For a good summary of the Big Bang theory and the evidence

for it, see P.G. Phillips, "The Thrice-Supported Big Bang," *Perspectives on Science and Christian Faith* 57 (June 2005): 82. See also Hugh Ross, *The Fingerprint of God*.

12. R.C. Sproul, J. Gerstner, and A. Lindsley, *Classical Apologetics* (Grand Rapids: Academic Press, 1984), chapter 7.

13. I have previously addressed this subject in D Snoke, "The Apologetic Argument."

14. See, e.g., W.M. Tscharnuter, in *The Birth and Infancy of Stars*, R. Lucas, A. Omont, and R. Stora, ed. (Amsterdam: North Holland, 1985); and T.Ch. Mouschovias, in *Protostars and Planets*, T. Gehrels, ed. (Tucson: University of Arizona Press, 1979).

15. See, e.g., Robert F. DeHaan, "Do Phyletic Lineages Evolve from the Bottom Up or Develop from the Top Down?" *Perspectives on Science and Christian Faith* 50 (December 1998): 260–71.

Chapter 7 Interpreting Genesis 1 and 2

1. A "land-based observer" view of Genesis 1 has been argued by Bible scholar Gleason Archer as early as 1955, according to Hugh Ross, in *A Matter of Days*, 233.

2. For example, G.H. Pember, *Earth's Earliest Ages* (Revell, 1911), reprinted in several subsequent editions.

3. C. S. Lewis, *The Magician's Nephew* (New York: Macmillan, 1955), 98–113.

4. For example, Paul H. Seely, "The First Four Days of Genesis in Concordantist Theory and in Biblical Context," *Perspectives on Science and Christian Faith* 49 (June 1997): 85–95.

5. St. Augustine, *The Literal Meaning of Genesis, vol. 1*, J.H. Hammand, trans., (New York: Newman Press, 1982).

6. As discussed by M.D. Futato, in "Because It Had Rained: A Study of Gen 2:5–7 with Implications for Gen 2:4–25 and Gen 1:1–2:3," *Westminster Theological Journal* 60 (1998): 1. The terms translated here "bush of the field" and "small plant of the field" are specific terms for Middle Eastern field vegetation.

7. Carol Hill, "The Garden of Eden: A Modern Landscape," *Perspectives on Science and Christian Faith* 52 (March 2000): 31–46.

Chapter 8 The Flood of Noah

1. See, e.g., Thomas Key, "Does the Canopy Theory Hold Water?" *Perspectives on Science and Christian Faith* 37 (December 1985): 223; Stanley Rice, "Botanical and Ecological Objections to a Preflood Water Canopy," *Perspectives on Science and Christian Faith* 37 (December 1985): 225; David F. Siemens Jr., "Some Relatively Non-Technical Problems with Flood Geology," *Perspectives on Science and Christian Faith* 44 (September 1992): 169–74; David F. Siemens, Jr., "More Problems with Flood Geology," *Perspectives on Science and Christian Faith* 44 (December 1992): 231; W.F. Tanner, "How Many Trees Did Noah Take on the Ark?" *Perspectives on Science and Christian Faith* 47 (December 1995): 260–63.

2. Verses 7:19–20 are taken from a direct literal translation (see appendix), not the ESV.

Chapter 9 Implications for Theology

1. Snoke, "Toward a Unified View of Science and Theology," 166–73; and "The Problem of the Absolute in Evidential Epistemology," 2–22.
2. For a good discussion of the theological premises involved in much evolutionary reasoning, see Hunter, *Darwin's God*.
3. Stephen Jay Gould, *The Panda's Thumb* (New York: Norton, 1980).
4. R. Dawkins, *The Blind Watchmaker* (New York: Norton, 1986).
5. Jerry Bergman, "Is the Inverted Human Eye Poor Design?" *Perspectives on Science and Christian Faith* 52 (March 2000): 18.
6. H. Endo, D. Yamigiwa, Y. Hayashi, K. Koie, Y. Yamaya, and J. Kimura, "Role of the Giant Panda's 'Pseudo-Thumb,'" *Nature*, 397 (1999): 309.

Appendix

1. The word for man here, adam, has the same root as the word adama, which means "dirt" or "red clay." Here adam will be translated as "human" and adama will be translated as "humus," to emphasize the connection. The two English words have the same Latin root, which, as in Hebrew, refers to tillable soil. The word adam does not have at its root the concept of maleness. Another Hebrew word, ish, means man, or male, in contrast to isha, "woman."
2. The name "Cain," here translated "Spear," comes from the word for reed, which was also used for recording ownership deeds and sales.
3. These names are not translated from the Hebrew, as the meaning is obscure.
4. These names are not translated from the Hebrew, as the meaning is obscure.

Subject Index

Abel and Cain, 65
Adam and Eve, 32, 49, 51, 56–57, 62–64, 68, 73, 74, 94, 97, 102, 152, 171, 178, 180, 192, 195
Alvarez, Luis, 41
angels, 58, 108–110, 177
apologetics, 123
apostles, modern, 16–18, 21
apparent-age hypothesis, 29–33, 47, 174, 185, 191
appearances vs. essences, 15, 21, 174
Aquinas, 122
archaeology, 116, 118, 146
argument from design, 28–29
atheism, 96, 123, 128
Augustine, 69, 72, 111, 140, 180

Babel/Babylon, 169, 180
balance of nature, 71, 76–79, 92–93, 97, 170–171, 178, 183, 193
Bayesian argument, 124
Behe, Michael, 44
Bergman, Jerry, 184
Big Bang, 127–129, 139, 148, 160
birth pangs, 73, 110

Calvin, John, 14, 122
Cambrian explosion, 129, 148, 160

canopy theory, 162, 186
carnivorous animals, 49–51, 58, 68, 75, 81, 92, 98, 177, 178–179, 183
circumcision, 66
coal and oil industry, 42, 46, 163
cold fusion, 187–188
concordantist science, 114–115, 121–122, 124, 127–130, 135, 181
continental drift, 37, 42, 160, 186
Copernicus, 14, 22
Coriolis force, 16
covenant of law, 57, 89, 95, 177, 178
covenantal theology, 176
curse, 49–51, 55, 57, 72, 94, 165, 171–172, 178, 184

darkness, 59, 86–89, 179
Darwin, Charles, 97, 118
Darwinism, 44, 96–97, 164, 184–185, 187
David, as type of Christ, 54
"day", 48, 100, 141–145
day-age interpretation, 100, 105, 111–112, 115, 121, 130
death, 57, 61–64, 74, 94
demons, 58, 134
destruction of Jerusalem, 18, 19
dinosaurs, 84–85, 148

Subject Index

Dirac, Paul, 28
dispensationalism 176
DNA, 11
Doppler effect, 27
Drosnin, Michael, 43

earth, motion of, 14–16
earthquakes, 39, 164
Eckelmann, Herman, 105, 115
electrons and protons, 11
"evening and morning", 48, 143–144
evidentialism, 123–124
evil, natural, 91
evolution, 43–45, 164, 183, 185, 193
ex nihilo creation, 91
existentialism, 118
"expanse", 140
extra-biblical texts, 19

fall of Adam and Eve, 48, 51, 65, 93, 184
fine-tuning of constants of nature, 28, 186
Fitzgerald, F. Scott, 188
flood geology, 39, 41, 155, 158–165
fossils, 40, 48
Foucault's pendulum, 14
framework interpretation, 100, 104–106, 113
Freud, 118, 119
futility of nature, 70–74

Galileo, 13–14, 15–16, 181
gap hypothesis, 134
garden of Eden, 53–56, 58, 63, 70, 116, 152–155, 157, 172, 180
geological sediment layers, 33–36, 39–41, 155
Gill, John, 86
glory of humankind, future, 70
Gnosticism, 117
God of the gaps, 121–124
Great Barrier Reef, 33
green plants, grass, 65–67, 68, 143, 152–153

heaven, 52–55, 69–70, 97, 179–180

hell, 56–57, 89
Hill, Caroline, 154
homeopathy, 190
Horton, Michael, 7
Hubble shift, 25
human brain, 130

image of God, 194
intelligent design movement, 13, 44, 119, 122, 124, 183–191

Jesus, Messiah, 52, 54, 56, 64, 66, 74, 102, 103, 110, 126, 169, 182, 186, 192
Johnson, Philip, 44
Joshua, prayer to stop the sun, 14, 15
Jubilee, 100, 113

Kepler, 22
Kline, Meredith, 7, 104
Kuhnian revolution, 124

"land", 133, 135, 152–153, 155, 166, 169, 175, 197
leap of faith, 118
leviathan/great sea monster, 59, 82–86, 147, 177, 180
Levitical dietary laws, 51, 116, 131
Lewis, C. S., 136–138, 179
Lindsey, Hal, 43
local flood hypothesis, 158, 168–171
Luther, Martin, 14, 71–72

MacArthur, John, 7, 141
magnetic stripes on ocean floor, 36–38
Maxwell, James, 28
Maxwell's equations, 28, 30
meteor impact, 41, 160, 171
methodological naturalism, 125
miracles, 112, 118, 124–130, 158–159, 181, 193
Morris, Henry, 43, 175

natural evil, 91
New Age philosophy, 178

219

Newman, Robert, 105, 135
Newton, Isaac, 26
Noah, 65, 88, 151, 158, 164, 169, 172–173, 182, 193, 195
nuclear physics, 27, 187

overpopulation, 65

Paley, William, 122
panda's thumb argument, 183–184
parallax, 25, 26, 27
parallelism, in Hebrew Scripture, 83, 105, 113
parasites, 96, 183, 184
Passover, 66
Peleg, 164
pollution, 73
postmodernism, 118
Presbyterian Church in America, 7

Reformation, 22
resurrection of Christ, 119–121, 131, 186
Roberts, Oral, 18
Ross, Hugh, 105, 115, 128, 135

Sabbath day, 99–100, 102–103, 150
Sabbath year, 101, 109
sacrifice, 65–66
Satan, 51, 180
Schaeffer, Francis, 8, 31, 116, 181
Schroeder, Gerald, 115
science, definition, 13
scientific creationism, 30, 31–32
sea, 59–61, 78, 81, 179
second coming of Christ, 18, 21, 23, 31, 52, 111, 174, 181, 194

second law of thermodynamics, 74, 178
Shapiro, Robert, 44
signs and wonders, 17, 23
snake, 51, 180
Sodom, 169, 174
solipsism (see also apparent-age hypothesis), 191
spectral analysis, 27
speed of light, hypothesis of changing, 28
spontaneous generation theory, 123
Sproul, R. C., 8, 56, 128
stars, distance to, 24–25
Stoner, Don, 33
sun and moon, 53, 105, 115, 147

temple of Jerusalem, 54–55
time, 32, 74
transsubstantiation view, 31, 46
Tree of Life, 53, 63, 177
typology, 53–55, 113, 172, 174

uniformitarianism, 38, 41, 161

van Til, Cornelius, 8
van Till, Howard, 125
variable stars, 25
vegetarianism, 64–68, 172, 178

Wesley, John, 14
Wiester, John, 115
wrath of God, 58, 93–96, 110, 179, 194

yin and yang philosophy, 76–77

Scripture Index

Genesis

1	45, 49, 56, 59, 68, 71, 77–79, 102, 106, 108, 113, 115, 117, 127, 130, 138, 156, 173, 177	
1–2	52, 59, 108, 152, 179	
1–3	53, 54	
1:1	133, 134	
1:1–2	135	
1:1–5	47	
1:2	134–135, 139	
1:3–4	91	
1:3–13	138–139	
1:4–5	78	
1:6	88	
1:6–7	85	
1:7	78	
1:9–10	78	
1:11–12	148	
1:14	162	
1:14–25	146	
1:16	147	
1:16–18	79	
1:21	59, 82, 83, 98, 148	
1:24	148	
1:25	147	
1:26–31	148–149	
1:27	62, 79, 151	
1:28	58, 68, 172	
1:28–30	66, 67, 102	
1:29–30	66	
1:30	65	
1:31	71, 75	
2	102, 152, 157, 180	
2:1	86, 165	
2:1–3	150	
2:4	141, 150, 152, 156	
2:5–7	152	
2:8	154	
2:8–14	116	
2:8–15	153–154	
2:17	57, 62	
2:19	147	
2:23	102	
3	165	
3:14–24	49–51, 56, 58	
3:16	73	
3:17	171, 172	
3:21	65	
3:22	63, 74	
4:4	65	
5	151	
5:1–2	151	
6–7	166–168	
6:1–13	166–167	
6:9	151	
6:17–7:12	167–168	
7–9	175	
7:11	142	
7:17–24	168	
7:19–20	165	
7:20	166, 175	
8:21	170, 171	
8:22	77, 170	
9:1–3	66	
9:1–4	64–65	
9:3	65, 66, 172	
9:11	171	
9:15	171	
10:1	151	
10:25	164	
11	169	
11:10	151	
11:27	151	
25:12	151	
25:19	151	
32:31	117	
36:1	152	
37:2	152	
41:57	166	
49:26	45	
49:27	144	

Exodus

10:21–29	86
14:26–15:8	86
15:19	60
20	60
20:8–11	99–100
31:17	100

Scripture Index

Leviticus
25   100–101
25:2–11   100–101
25:4   101

Numbers
14:32–33   19
21:8–9   54
29   145
32:29   67

Deuteronomy
5   66
28:15–68   57
32:24   86
32:33   84
33:15   45

Joshua
10:12–13   14
18:1   67

1 Samuel
8:8   141
8:18   141

2 Samuel
22:19   141

Job
4:20   144
7:12   60, 84
11:7   95
15:31   72
38   81
38–41   80, 98, 177
38:4   81
38:4–7   138
38:4–23   79–80
38:6   81
38:8–11   80, 170
38:12–13   81
38:16–17   81
38:19–20   81
38:22–23   81
38:26–27   81
38:34–35   81
38:39–41   80, 81
39–41   82
39:3   82
39:6   82
39:16   82
39:20   82
39:30   82
41   83, 84
41:14   82
41:34   84

Psalms
2:7–8   15
8:4–6   58
30:4–5   144
32:1   15
49:12   63, 92
49:14   144
49:20   63
74:13–14   82, 85
74:13–17   85
74:14   85
74:15   85
74:16   85
74:17   85
74:19   86
78:53   60
90:2–6   143
90:4   100
90:10   144
90:13–14   144
91:20   92
93:1   14, 15, 22, 23
93:4   60
96:10   14
104   88–89, 98, 173, 177
104:1–32   86–89
104:2   88
104:2–5   88
104:5   14
104:6   88
104:6–8   88
104:6–9   170
104:9   88
104:11–14   88
104:19–23   88
104:20   88
104:21   88
104:23–29   60
104:25   88
104:27–30   88
145:3   95

Proverbs
3:19–20   85
8:27–28   140
8:27–29   80
20:10   189

Ecclesiastes
1:5–7   71
1:8   72
3:11   95
3:18–20   63
11:5   95
11:6   144

Isaiah
2:20   141
3:7   141
3:18   141
5:30   60
7:21   141
11:6–8   52
17:12   60
21:12   144
27:1   82
28:15–18   178
34   55
34:4   52
40:17   72
40:28   95
45   173, 179
45:5–12   90–91
45:7   91, 98
57:20   60
66:22–24   56

Jeremiah
4:23–28   134–135
5:22   60, 80, 170
5:22–24   92–93
10:10–13   93
10:13   81
12:9   86
50:42   60
51:34   84
51:42   60

Ezekiel
5:17   86
11:19   62
14:15   86
14:21   86
26:3   61
27:32   61
29:3   84
36:26   62

Daniel
5:1   19
7:2–3   61
7:5   84
9:24   17

Amos
9:3   61

Jonah
1:15   61
4:8   117

Zechariah
3:9–10   142

222

Scripture Index

3:10 156
10:11 61

Malachi
4:5–6 19

Matthew
5:3–12 66
6:26 68
8:12 55
8:22 62
10:1 17
12:39 17
13:52 195
16:24–28 18, 23
21:33–39 17
21:33–43 19
22:13 55
24:4–8 110
24:34–35 17
24:37–39 169
25:30 55

Mark
8:34–9:1 18

Luke
6:20–26 66
9:23–27 18
17:26–27 169, 172
17:27 170
17:28–29 169
17:29 170
21:20–24 19
21:25 61

John
5:17 102–103
6:13 159
9:18 159
16:12–13 17

Acts
17:6 169
28:4 61

Romans
1:8 94
1:18 94
1:18–20 94
1:20 57, 89, 94, 95
5 192
5:12 62
5:12–15 61–62
5:12–19 64
5:15–17 192
5:18 64
8:19–22 69, 72
8:20 71
8:22 73
14:2 68

1 Corinthians
6:3 58
9:22 191
12:28 16
13:8 17
15:3–8 120
15:5–8 159
15:6 178
15:13–17 120

2 Corinthians
10:5 191
11:26 72
12:12 16, 23

Ephesians
2:1 62
4:11 16, 23

1 Timothy
4:4 71, 72, 75

2 Timothy
1:12 182

Titus
1:15 71

Hebrews
1:1–2 17
2:7 58
2:9 70
4 103
4:3 103, 107
4:3–4 103, 112
4:4 103
8:1–5 54, 55

1 Peter
3:20 170

2 Peter
2:5 169
3 175
3:4 174
3:5–7 173
3:6 169, 174

3:7 174
3:8 100, 142

Jude
13 61

Revelation
1–19 194
5–6 107
6:8 86
6:12–14 109
6:13–14 52
6:17 110
8–9 107
8:1 107
8:7–12 110
8:13 109
9:10–11 109
9:12 110
9:15 110
9:19 109
11:14 110
11:15–18 108
13:1 61
15:1 110
16 107
16:1 108
16:8–10 110
16:10 86
18:21 86
19:1–3 56
21 180
21–22 52–55, 194
21:1 53, 59
21:5 52
21:23 53
22:4–5 53

223

David Snoke (Ph.D. 1990, University of Illinois at Urbana-Champaign, B.A. magna cum laude 1983, Cornell University) is an Associate Professor in the Department of Physics and Astronomy of the University of Pittsburgh. From 1993 to 1994 he was a staff physicist at the Aerospace Corporation in El Segundo, California, and from 1990 to 1992 he worked at the Max Planck Institute in Stuttgart, Germany, as an Alexander von Humboldt fellow. He has published more than seventy scientific articles in international journals and two scientific books for Cambridge University Press on his work on lasers and semiconductors, as well as several articles on Christianity and science for *World* magazine and for *Perspectives on Science and Christian Faith*, the journal of the American Scientific Affiliation, and a high school physics textbook that integrates Christian theology, philosophy, and physics. He recently co-authored an article in *Protein Science* with Michael Behe, which is one of the few recent papers in a refereed science journal to explicitly express skepticism about Darwinian evolution. He is licensed to preach and an elder in the City Reformed Presbyterian Church, in Pittsburgh, Pennsylvania, which belongs to the Presbyterian Church in America denomination. He and his wife, Sandra, have four children, whom they have homeschooled.